When the Forest Breathes

When the Forest Breathes
Renewal and Resilience in the Natural World

SUZANNE SIMARD

ALLEN LANE
an imprint of
PENGUIN BOOKS

ALLEN LANE

UK | USA | Canada | Ireland | Australia
India | New Zealand | South Africa

Allen Lane is part of the Penguin Random House group of companies
whose addresses can be found at global.penguinrandomhouse.com

Penguin Random House UK
One Embassy Gardens, 8 Viaduct Gardens, London SW11 7BW

penguin.co.uk

First published in the USA by Alfred A. Knopf 2026
First published in Great Britan by Allen Lane 2026
001

Copyright © Suzanne Simard, 2026

The moral right of the author has been asserted

Page 311 constitutes an extension of this copyright page

Penguin Random House values and supports copyright.
Copyright fuels creativity, encourages diverse voices, promotes freedom
of expression and supports a vibrant culture. Thank you for purchasing
an authorized edition of this book and for respecting intellectual property
laws by not reproducing, scanning or distributing any part of it by any
means without permission. You are supporting authors and enabling
Penguin Random House to continue to publish books for everyone.
No part of this book may be used or reproduced in any manner for the
purpose of training artificial intelligence technologies or systems. In accordance
with Article 4(3) of the DSM Directive 2019/790, Penguin Random House
expressly reserves this work from the text and data mining exception.

Printed and bound in Great Britain by Clays Ltd, Elcograf S.p.A.

The authorized representative in the EEA is Penguin Random House Ireland,
Morrison Chambers, 32 Nassau Street, Dublin D02 YH68

A CIP catalogue record for this book is available from the British Library

ISBN: 978–0–241–76331–5

Penguin Random House is committed to a sustainable future
for our business, our readers and our planet. This book is made from
Forest Stewardship Council® certified paper.

With love to my parents,

ELLEN JUNE SIMARD (NÉE GARDNER)
(1936–2022)

and

(PETER) ERNEST CHARLES SIMARD
(1935–2024)

What we do to the land, we do to ourselves.

—LISA WHITE-KUUYANG, *Haida Nation*

CONTENTS

Map	xi
Author's Note	xiii
INTRODUCTION: WHEN THE FOREST BREATHES	3
1. CAN'T GET AWAY FROM YOUR ROOTS	13
2. KINSHIP	28
3. PLANTING SEASON	48
4. KILLER FORESTS	64
5. GARDENS IN THE FOREST	79
6. THE MIRACLE OF YEW	95
7. SISTER CEDARS	107
8. CARBON BOMB	125
9. PLACE MATTERS	144
10. RECIPROCITY	157
11. THE FOREST DEFENDERS	169
12. THE FIRST ECOLOGISTS	182
13. IN THE ARC OF A TURN	194
14. TREES IN OUR BLOOD	203
15. BACK TO THE GARDEN	218
CONCLUSION: FINDING OUR WAY FORWARD	231
Acknowledgments	239
List of Species	245
Notes	251
Critical Sources	255
Index	293

This map shows the Mother Tree Project sites; the nearby towns, rivers, and mountain ranges; and the names of only those First Nations connected to the individuals or locations mentioned in the book. There are more than 2o3 First Nations recognized in British Columbia by Canada's Indian Act, with some First Nations still fighting for their rightful recognition. This map is only an approximation and not meant to be a comprehensive or accurate representation of all of the vast and often overlapping Indigenous territories.

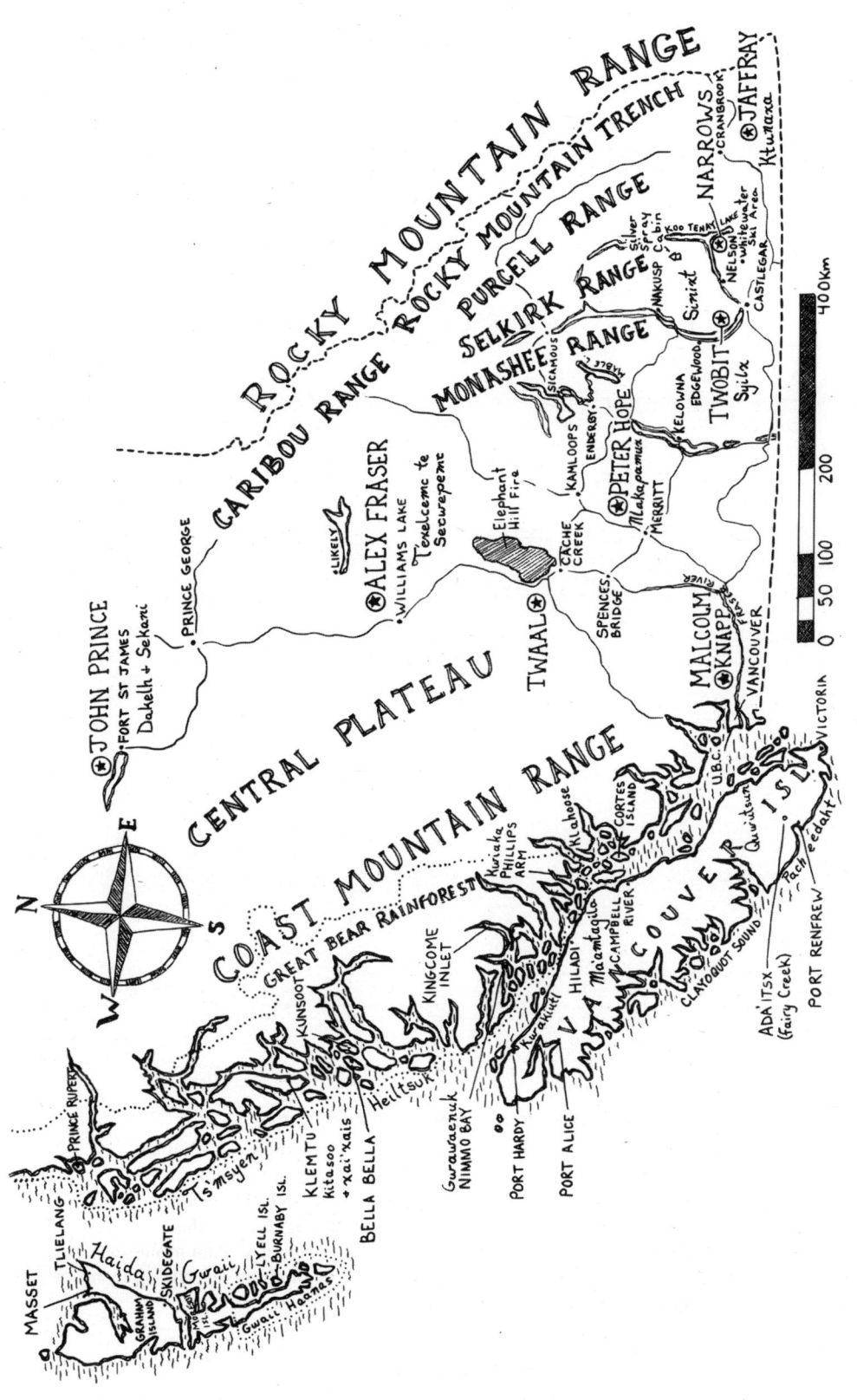

AUTHOR'S NOTE

This book builds upon my 2021 memoir, *Finding the Mother Tree: Discovering the Wisdom of the Forest.* In these pages, I weave my recent research in the forests of British Columbia with my lived experiences from 2015 to 2023. I have conducted this research, alongside my students and colleagues, in my position as professor of forest ecology in the Faculty of Forestry and Environmental Stewardship at the University of British Columbia. I have interpreted my work in ways I feel may help the public grapple with complex concepts regarding ecological resilience, and that may also motivate improved stewardship of the forests of British Columbia. The forests in which I conducted this work include the inland and coastal temperate forests of British Columbia. This writing is from my perspective, developed from my relationships with the land, its people, and my ancestors. My ideas represent the confluence of my worldview, work, and life, and others may not agree with me. I am only one voice of many, and many voices must be heard.

While conducting my research, I had the deep honor of working alongside Indigenous people who generously shared with me their philosophies, knowledge, and culture with respect to the forests of the Pacific Rim. This book includes reference to Indigenous knowledge, but it is important to understand that this knowledge cannot be taken, misused, or exploited. This knowledge comes from learning from the land through a cultural lens and cannot be replicated or duplicated in the absence of the people or place from where it came. Any references I make to specific practices are meant only as examples to illustrate a broader philosophy of care and stewardship of the land. It is my hope and intention that my work shines a light on the truth of the colonial exploitation of the forests of Canada, and hence of Indigenous land and people, and that seeing, believing, and acting on this truth leads toward respecting, tending, and healing of the land and its people.

At the end of this book, in the section "Critical Sources," I list the peer-reviewed research articles, by chapter, that informed my writing, including articles originating from my lab. I also include other papers that influenced my thinking or that I refer to directly. I invite readers to consult with these sources so they may delve more deeply or verify the findings if they wish. The book is not meant to be read or critiqued like a peer-reviewed scientific research paper. This writing is simply my interpretation of the work I have been involved with in the past decade.

For names of species in this book, I use a mixture of Latin and common names throughout. A comprehensive list of Latin and common names of all species mentioned are included at the end of the book. In developing the chapters, I often integrate several experiences into a single story in order to provide a good pace and flow for readers. In writing this memoir, I have drawn from a mix of experiences, notes, writings, conversations, interviews, and readings. I have changed the names of some people to protect their privacy. Any faults in my research, errors, omissions, or misinterpretations of any research, stories, or histories are my own. I wrote this book with what I knew to be true at the time of this writing.

When the Forest Breathes

INTRODUCTION

There is no way to understand how something grows without also understanding how it dies. When you walk in the forest, you see the abundance and growth, but if you look closer, you will also see disturbance and decay. There is beauty in life, but there is also beauty in death and in what remains behind. In the forest, renewal only comes with dying. It is part of my life's work to understand how this renewal happens and what these processes have to teach us.

Nature is a rich fabric of endless cycles, from mushrooms and soil animals breaking down logs, to bacteria and fungi transforming organic sources and creating "dead" matter that is full of life, to aging trees "downloading" their genetic knowledge to their young. As young trees mature, they learn, adapt, and grow with their evolving circumstances and pass their wisdom to the next generations, ensuring the forest is enduring, regenerative, and resilient. Our own lives are inextricably bound by the same rules and destiny: birth, growth, death, decay, rebirth.

Nature is waiting for us to listen. And to learn.

THESE TEACHINGS WERE brought home to me in the aftermath of my own encounter with death. Decades of study in the inland forests of my home province of British Columbia had taught me that the web of community—the trees and plants, humans and animals, fungi and microbes, and the breathing soil itself—exists in a vibrant dance, a clockwork of adaptive cycles. Even as I marveled at the way energy cycles through this web, my own life energy was unraveling inside of me. I wrote my first book while recovering from breast cancer. I leaned on the teachings of the old trees to help me heal. In turn, I was able to use the insights I'd learned—about how connection, community, and wisdom are vital to recovery—to ask deeper questions about the forest.

When *Finding the Mother Tree* was published in 2021, I was astonished by the public's heartfelt embrace of my scientific discoveries. Reader after reader wrote to thank me for confirming something they said they'd always known in their hearts—that the forest was a connected, intelligent, living system. One of my key findings was that the biggest, oldest trees—the mother trees—are the energetic keystones of the forest, the hubs of an underground network of mycorrhizal fungi connecting the forest. I used the metaphor of the mother tree because of their role in dispersing seed to the understory, connecting with the tender roots of their offspring, and nurturing the new shoots. Mycorrhizal fungi form symbiotic relationships with tree roots and trade vital nutrients gathered from the soil in exchange for photosynthetic carbon from the trees. A symbiotic relationship is one that is close, cooperative, and interdependent, the result of coevolution over millennia to ensure fitness of the members. These fungi can connect the trees and seedlings together in an extensive subterranean web and serve as pathways for shuttling carbon, nutrients, and water through the forest. The connections and communication provide the ecosystem with strength and resilience.

Generations of trees and plants are deeply integrated in this system of interactions and feedbacks, and these relationships are crucial in the regenerative capacity of the forest. My colleagues and I found that the belowground linkages had patterns suggestive of a biological neural network, that the trees have agency in the transmission of resources, and that together, these properties are suggestive of a type of intelligence. Even more intriguing, the trees demonstrate traits such as collaboration and facilitation, not just dominance and competitiveness. Life in the forest, it turns out, is more complex than "survival of the fittest." An intricate network of cooperation and competition makes the ecosystem thrive.

The forest wasn't my only teacher. I learned from my Indigenous colleagues that First Peoples have always recognized the interconnectedness of forests and of all living things. Many Indigenous cultures have long viewed mother trees—and grandmother and grandfather trees—not just as metaphors for interdependence, but as sacred relations who hold spiritual connection to the ancestors, and who provide wisdom, longevity, and regeneration. But Indigenous ways of seeing the world are rarely accepted in Western science methodologies. They are in fact roundly disparaged. Even the use of metaphor, such as "mother tree," in Western scientific language has been criticized as lacking objectivity and integrity. I found myself longing to push back against these rigid boundaries of science, to learn additional ways of inquiry, other ways of seeing and knowing the natural world. And I had new questions, different questions.

So, early one spring morning, I went back to the forest. I wanted to learn what the mother trees could teach me about survival and adaptability, about living and dying, and about how energy and ancient carbon cycle through the forest. I wanted to find out how the forests were coping with stressors like climate change and clearcut logging—the practice of harvesting every tree in a large expanse of forest. I had a hunch that the forest's connections, cycles, and diversity could guide us to the solutions we so desperately needed to enhance resilience and protect our global ecosystem.

Hiking deep into the woods, I stopped to rest at a shady hollow, a secluded hiding place for shade-loving trees and plants. I admired a small cluster of fairy slippers, exquisite purple orchids whose lives depend on mycorrhizal fungal connections in the soil. I sat on a decaying log, an old mother tree that had long since passed. Her great body was sprinkled with seedlings of western hemlock that had taken root in her softened bark. I marveled at how the old tree had created a rich environment for new life to grow, cycling back to a beginning.

As trees weaken, perish, and collapse, they change from living biomass to necromass. The soil food web—the community of organisms including animals, fungi, and bacteria—breaks down the rotting logs and decaying litter while extracting carbon energy. Through the complex process of decomposition, nutrients are released and carbon dioxide is respired to the atmosphere. Seedlings establish in the carcasses of old mother trees and unfurl their delicate young leaves to absorb the carbon dioxide, while slurping up the nutrient soup with their tender mycorrhizal roots. The soil and plant community continue the cycle of regeneration as the biomass becomes deeply *alive* and enriched. The whole forest awakens in a teeming, bustling surge, as rich and joyful as the young at play and the old at work, carrying forward the business of life.

Cavity nesters like woodpeckers, flickers, squirrels, and even bears build nests in the soft decay of the elders and reproduce in the township of the dead and dying. Animals and arthropods, centipedes and springtails disperse the immortal remains of hollowed trunks and broken branches, feeding extensive food chains through the extended ecosystem and creating structures for the living. Carbon compounds seep into the full life of the subterranean ecosystem and recombine into organo-mineral compounds that remain stable in the soil. In an intact forest, this carbon pool—the detritus of plant, animal, and microbial life—is safeguarded in the soil for millennia. The compounds of the old are transmuted into the living, cycling onward for centuries to come.

Yet many of our current forestry practices harm these extraordinary natural cycles. Industrial clearcutting has removed vast amounts of forest, and recovery of these ecosystems is slow. The massive machines not only cut down trees, but disrupt the fragile connections of the forest, both above and belowground. It can take centuries for a forest to recover, millennia for the soil to rebuild.

As I pondered this, I wondered if there could be better ways to tend forests that would protect their cycles, connections, and niches, and the biodiversity they support. Given that people will continue harvesting trees or parts of trees for buildings, paper, clothing, and many other products, surely we could find gentler ways to do it, and take far less in the process. I tried to envision a silviculture system—a program to harvest, regenerate, and care for a forest—that would actually leverage the forest's natural regenerative capacity and biodiversity. One that would preserve and nurture the mother trees. We needed to protect these legacies, and in so doing, alleviate the stresses of changing climate, resist or dampen wildfires, and stave off insect attacks and pathogen infestations. Fostering abundance and balance in the plant and animal communities could potentially enhance carbon sequestration and storage, practices that held promise for mitigating climate change. I knew that the mother trees were the powerhouses of the forest. If we could conserve these big old trees, they could naturally seed the sites, regenerate the forest, and help buffer the ecosystem from stressful conditions.

What I was imagining would require a radical change—a transformation—in the direction of the forest industry in British Columbia, but surely it was worth pursuing.

Of course, the idea of humans tending forests in this way is nothing new. Forests have been shaped by people since the beginning of humankind. For thousands of years, the First Peoples cared for the forests and waters by aligning with the natural processes that fostered abundance for generations to come. They developed sophisticated practices such as strategic burns that increased food production, enhanced plant diversity, improved the soil, and prevented catastrophic wildfires. When Europeans crossed the Atlantic to North America, they saw only inexhaustible wealth in the vast forests and wild rivers and the plentiful species that inhabited them. They did not grasp that the bounty of resources was deeply influenced by age-old ancestral stewardship that honored the natural cycles and connections of the ecosystems.

The First Peoples' knowledge and governance systems are based in respect, reciprocity, and responsibility for the nurturing of living beings. The silviculture system I hoped to apply would be guided by these traditional perspectives and a holistic understanding of the natural world. It would embrace the wisdom and intelligence of the forest and employ different ways—more adaptive ways—to protect and care for forests in our changing climate.

My determination to find a better way forward was the catalyst for the Mother Tree Project, a vast study that would take me into the heart of the Pacific Rim forests. My goals were to document the effects of clearcut logging compared with partial cutting, or just leaving the forest to its own ingenuity, on carbon pools and biodiversity—two elements our lives depend on that are rapidly changing as human populations grow and the climate warms—to help us discover new ways of tending forests that would protect the mother trees. I wrote the proposal with my postdoc Brian Pickles and my close colleagues Les Lavkulich and Bill Mohn and others, and much to our surprise, it was funded by the federal government of Canada. It would become one of the biggest forestry field experiments in Canada, with sites distributed halfway across British Columbia, and involve collaboration with many allies, including Indigenous people, local communities, academics, governments, logging companies, artists, and students.

Jean, my longtime colleague and best friend of more than forty years, took on the role of managing this complex project. With the help of Brian, she located nine experimental Douglas fir forests, eight of them in the interior of British Columbia where the winters are cold and the summers are warm, and one on the mild, drizzly coast east of Vancouver. Across this climatic range, our experiment would compare five different logging patterns, or treatments, to test how protecting different numbers and configurations of the largest overstory, or mother, trees would affect seedling regeneration, the understory plant community, bird and animal habitat, soil composition, and fire risk.

We sought the wisdom of scholars and managers who'd spent their careers investigating silviculture systems. These were experts who had done careful work learning how forests functioned and using the knowledge to develop sustainable management practices. But it wasn't until I engaged with postdoctoral fellow Teresa Ryan of the Ts'msyen (Tsimshian) Nation that I realized something essential was missing from our project.

Teresa and I had become fast friends and spent many hours discussing the social integration of her culture with the land and how it contributed to the vitality and balance of forest ecosystems. She described how Indigenous stewardship practices were in synchrony with the rhythms of nature. How salmon populations were tended in alignment with tidal cycles, which nourished the forests, animals, people, and all relations. I learned how the ecology of forests had been deeply influenced by ancient stewardship practices, and most importantly, by the Indigenous worldview that all is connected. In turn, I could see how profoundly the people, and their origin stories, had been shaped by the forests. Forests I once naïvely thought were mainly the product of climate, soil, and the lay of the land, I now understood were reflections of the philosophies and practices of the original people.

Until that moment I had spent my scientific career trying to improve upon colonial forest stewardship practices, striving to conserve an ecosystem's legacies and connections instead of harming them. Now I understood that not only were these forest practices misaligned, but the goals were wrong. We needed to transform away from a mentality of exploitation to a regenerative worldview that valued reciprocity, synergy, and abundance for the entire socio-ecological system.

My conversations with Teresa felt like a homecoming. I had finally found a system of stewardship my values aligned with, instead of chafed against. The colonial system desperately needed to integrate with the forest, in order to restore balance and stave off the losses in biodiversity and productivity it was causing through the commodification of the landscape.

The goals of the Mother Tree Project began to come into sharp focus. Teresa helped me and Jean connect with Indigenous people who knew the land best, as our nine sentinel forests were located within their territories. As much as we could, we discussed the project with the people and learned how to make adjustments for each place. We did our best to conduct our work with respect and humility, to provide principles and frameworks for stewardship of the land.

We held countless virtual meetings with our partners debating the details of the five experimental treatments, such as which specific dominant trees to leave standing. We already knew that protecting these big trees during logging would be important in helping forests regenerate, as happens following natural disturbances like wildfire. Extreme conditions were already intensifying with climate change, bringing unpredicted droughts, heat waves, downpours, and heavy snowfalls that easily killed young trees. If we could design forest practices that protected large trees, they could help mitigate these extremes while at the same time foster thrifty communities of trees, plants, and animals. Our investigations had shown these large trees interacted with their seedlings in profound ways, and this led us to naming the project "The Mother Tree Project."

After our nine research forests were logged in the patterns we'd specified, the Mother Tree Project crew—our group of dedicated students—and a team of professional tree planters planted mixtures of tree species among the remaining old trees so we could identify the combinations most resilient to the stresses wrought by the shifting climate. It would take us eight years before the last patch was logged and our final seedings established. By the time we were done, we had planted 225,000 trees across an area the size of Denmark, from the rugged interior of the province to the lush rainforests of the West Coast. We planted in the pouring rain and in the blistering heat, enduring tick bites, wasp stings, and close encounters with moose and bears. I visited old-growth forests that were pristine and unspoiled, and walked in clearcuts where every tree had been felled and the valuable ones taken by the lumber companies, leaving behind nothing but stumps and slash. We ran from a wildfire ripping through one of our experiments. I experienced the joy of watching my daughters fall more deeply in love with the forest as both Hannah and Nava pursued careers in forest ecology and joined me in fieldwork on the Mother Tree Project. Nava and I participated in the movement to protect the ancient forests of Fairy Creek, which became a flashpoint for logging practices in Canada.

We would return season after season, year after year, to measure, record, analyze, report, and archive what our nine forests were showing us as they responded, attuned, and adapted to our treatments and to the changing climate. We followed the seedlings from their establishment phase through juvenile and sapling stages, and made detailed plans for others to continue the work after we were gone, following the forests into their maturing and old-growth years, after the mother trees had died and spawned new generations. New professors and students would join us, examining in exquisite detail the responses of the wildlife, soils, hydrology, and microclimatic buffering to all of the treatments and climates.

Through all this, as I witnessed firsthand nature's cycles of death and renewal, I experienced parallel rhythms of growth, loss, and recovery in my own life. My children grew into saplings alongside the trees they planted, my work flourished and stumbled in tandem with the changes of our forests, my knowledge and spirit deepened as the ecosystems reorganized, and my mum moved through the final stages of her life as wildfires took the lives of some of the mother trees across our experiments.

In the pages that follow, I'll share these stories and some of the most remarkable findings from the Mother Tree Project. This huge study will reap discoveries for decades and will change and grow under the careful watch of those who come after me and Jean. It is my hope that this collaborative work will serve as an ecosystem from which to draw wisdom, providing insights on how to live attuned with the land and with one another, long after we are gone. Already we have made surprising discoveries directly related to how trees, from elders to seedlings, adapt to changing conditions. Plants and fungi and animals interact and collaborate, and disrupted forests work to seed new communities that are changed, transformed, and resilient in the shifting landscape. Because we humans are part of this community, we are an essential part of solutions already underway.

The Douglas fir forests depicted in this book are unique to their geographic place, as are all forest ecosystems. But the same basic life rhythms and cycles exist in all forests, whether the Pacific Rim coniferous forests, the Russian steppe, the Amazonian jungle, or the Mediterranean woodlands. All forests reflect the wonder of regeneration and healing, the ability to adapt, learn, and evolve with change, and to persist over decades, centuries, and millennia. In every cycle across the seasons, forests renew and restore. Trees germinate, grow, and set seed. They disperse their genes in the gusts of wind, feathers of birds, and gravity of freefall in a continuous adaptive cycle of birth, growth, death, decay, and rebirth.

We are part of this same cycle, inseparable from the trees, animals, water, and air. The forest teaches us to live in synchrony with the natural cycles, to stay in rhythm with the seasons, lifespans, and oscillations, the essential patterns that have evolved for success. I would carry my mum's life energy forward, just as my daughters and the generations to come would carry on my life energy, all of us constantly renewing, growing, maturing, and dying, spreading seeds of knowledge through the ages. The energy and wisdom of the old cycles evolve into the new. We and every other living thing, all our relations, are part of this infinite, beautiful, self-renewing cycle of existence. Even after the most destructive logging practices, forests strive to heal, although healing will be much faster if any logging is done with respect and care.

Nature is waiting for us to learn that we are all bound by the same rules. And to remember that, when the forest breathes out, we breathe in. When the forest thrives, we thrive. When the forest lives, we live.

1

CAN'T GET AWAY FROM YOUR ROOTS

It was late afternoon and the August sun was slanting crimson through the trunks of the Douglas firs. The forest felt eerily still. The trees seemed to flag. I heard a tree fall in the distance, and I shivered and looked over my shoulder. *Just tired,* I thought. Every inch of my body was aching from two solid weeks measuring every tree, log, and soil layer in this forest. I loved the work, but the heat and exertion were taking a toll.

We were working in the Cariboo Mountains, a short drive from Likely, British Columbia, and an eleven-hour commute from our hometown of Nelson, establishing the boundaries for one of the Mother Tree Project plots. If all went according to plan, this plot would be logged three months later, in November 2018, and replanted with our special mix of seedlings the following spring. I banged a foot-long piece of rebar into the ground with my axe, marking the center of the plot, and ran the end of the tape due north fifteen meters through the tangle of creamy honeysuckles and ivory snowberries. The warm, woody air drew streams of sweat through my hair that soaked my ball cap. A red-breasted nuthatch sang a high-pitched *neen neen neen,* as though playing a tin trumpet in rhythm with our work. I closed my eyes for a moment, feeling the embrace of the mountains. I could hear the ringing thump of Jean's axe as she established the other end smackdab south, completing the thirty-meter north–south transect.

For the past forty years, Jean and I had worked side by side in the bush, studying how forests responded to the many ways prospecting foresters had cut them down and tried to grow them back. In our early twenties, we'd scrambled up trees to escape grizzlies in the Lillooet Mountains and scaled the craggiest peaks carrying twenty-kilogram packs on our backs. We were bound by our shared love for the woods, by our triumphs and our stumbles. When my daughters were born, I'd plopped them down in the middle of lush thimbleberry bushes to play while we worked. "Can I eat this, Jean?" Hannah would ask, her chubby cheeks and dark hair already smeared with the scrumptious maroon berries. Now both daughters had grown lean and perceptive, their childhoods mostly a memory, while Jean and I were older and more experienced, and more resolved than ever to finding a solution to the ecological crisis that threatened us all.

Peering through the stand of Douglas firs, I could make out Hannah and Nava, eighteen and sixteen years old, briskly laying down a line perpendicular to ours. The heat didn't seem to bother the girls much. They were scaling over rotting logs, ducking under scarlet-

Suzanne, twenty-four, (left) and Jean Mather (later Roach), twenty-five, on Hong Kong's Peak. After earning our undergraduate degrees in forestry from the University of British Columbia in the spring of 1983, we backpacked through Asia for eight months on ten dollars a day. We ran out of water while hiking on Lantau Island and were drinking from cracks in the rocks. When we finally made it to a small abandoned village, we opened a barrel of water to find it was full of snakes. We eventually found mandarin orange juice at a street vendor, and Jean exclaimed, "I love drinks!" which became a standard joke whenever we found ourselves in a pickle.

stemmed Douglas maples, and threading the tape directly through plumb fir and feathery cedar saplings to ensure it was dead straight.

Beautiful, I thought. I was proud of my daughters.

At one end of the east–west tape, I spied Nava poking her graduated trowel into the ground every two meters to measure the depth and substance of the forest floor. Her nut-brown hair tied back with a shoelace and a sprig of willow, she pushed at the duff, gently separating the fibers and roots with the metal blade. My younger daughter was pensive and curious. I could hear her whispering thanks to Mother Earth for showing her these hidden secrets.

Hannah had started along the north–south line, calling out forest floor depths as Jean jotted down the numbers. At five foot two, Hannah was a good five inches shorter than Nava and just as fast, her legs lithe from leaping over creeks, arms muscled from packing field gear. Both girls knew just where the greasy black humus, the deepest layer of the forest floor, stopped and the gritty mineral soil started, signifying the depth to which organic carbon could easily be counted.

The two girls raced each other down the lines, their bodies agile, carrying the genes of French *scieurs de bois* from my side of the family dating back generations. The forests of Europe had provided for our family and prepared them for the woods of North America when our ancestors emigrated five centuries back. The carbon-rich trees and waters of the world's forests were knitted in the sisters' blood and bones. Even though they had planned to veer in different directions, one toward human health and the other to the orcas in the Salish Sea, the forest always called my daughters back. "Can't get away from your roots," my mum, Granny Junebug, would say, as they sat on her back porch sipping lemonade in the forest-clad valley of the Kootenays where our family had settled. Mum grew up on the edge of the woods in the dirty thirties, didn't suffer fools gladly, and knew how to get out of a pickle fast. "Git yer stuff and let's get the hell outta here!" she would say when confronted with a wildfire or an irate bear. She'd instilled in her granddaughters an instinct for survival and a passion for tending the soil and growing a garden, as well as the knowledge that they'd better get a darned good education if they valued their independence.

Ellen June Simard, aka Mum or Granny Junebug, seventy-seven, at Lakeside Park in Nelson, British Columbia, in Sinixt, Ktunaxa, and Syilx territories, 2013. Her faux leopard-skin pants were a reflection of her free spirit, which she joyfully exercised every day, whether chugging over the mountain passes in her VW van to go windsurfing, joining land defenders protesting the logging of old-growth forests, or zooming down mogul runs at Whitewater.

After ten minutes, the four of us had established the two line transects, each pinned at the middle to the center point, forming a cross in the forest. The ends demarcated the quarter chimes of the circular plot, which looked half the size of a hockey rink. Next, we would methodically record every detail of the plot: the large trees, the small trees, the plant community, the woody debris on the ground, the forest floor, and the layers of mineral soil beneath.

I glanced over at the forest within the plot, seeing trees young and old, shrubs laden with berries purple and red, paintbrushes and arnicas glowing scarlet and yellow. Gaps had opened where some trees had fallen, allowing new light to reach the leaves of willows and alders. Old trees at the edges of the gaps had produced a multitude of cones, spawning new seedlings that were enlivened by the shafts of sunlight. Large decaying logs on the forest floor stored carbon and water while providing homes for shrews and salamanders as well as new conifer seedlings. In the duff, chocolate brown humus was cycling carbon and nutrients to deeper, more protected horizons beneath. The bed of an ephemeral stream meandered through the center. A snag shedding graying bark stood leaning, a pileated woodpecker foraging its cambium for insects and sapwells.

I envisioned layer upon layer of stomata, the very lungs of the trees, in this highly structured woodland, with foliage folded under and over itself in brilliant strata from the forest floor to the treetops. Life felt capable here. Solutions were wired into this place. Our experimental plot, which was split into four segments, aimed to capture the teachings of a medicine wheel, a symbol that has been used for generations by Indigenous cultures around the world. Among its many interpretations, the medicine wheel embodies a worldview of interconnection and the cyclical nature of life: birth, youth, elder, death. Air, water, earth, and fire. Winter, spring, summer, fall. Anyone with a metal detector would be able to find the center pin of this plot, the heart of the medicine wheel, even if the forest burned and found new life again.

Jean and I got busy recording details about the trees while Hannah and Nava turned to measuring downed wood along the transects. Sweat trickled down my spine as I squinted through my laser at the leader of the tallest Douglas fir and shouted its height: "Forty-two point six meters!" This tree was a good two meters taller than the green shoots of its neighbors, and its girth thicker. Jean measured its diameter at 1.3 meters above the ground. The chunky, corky bark was scarred on the north side from a ground fire a century ago, and on the east side by another fire decades earlier. I imagined its massive roots tapping into underground rivulets of water, enabling it to resist the flames. I heard Jean cranking the increment borer, then calling the tree's age. "Two hundred and fifty-three!"

"Is it a mother tree, Mum?" Hannah glanced up from her meter-long calipers clutching a rotten log.

"More like a *grand*mother." This tree was not only a giant, she was the oldest in the patch.

I shifted my laser to measure the height of another elder tree as a black-and-yellow swallowtail butterfly fluttered down through the branches, then wafted up again in the soft, warm air. A mirage of carbon dioxide and water molecules from photosynthesis and transpiration was hovering in the downdrafts and updrafts. I blinked, paused to take a drink of water, looked again. The molecules were splitting into their bare elements: carbon, hydrogen, and oxygen. Carbon, with six each of protons, neutrons, and electrons, was forming covalent bonds and joining into long-chain organic polymers. Then it was flipping between its two stable states, ^{12}C and ^{13}C, and its natural radioactive species, ^{14}C. Forming graphite and diamonds, right before my eyes. *Heat will do that to a person,* I thought.

Carbon is the element of beauty, essential to life on Earth. Carbon dioxide is fixed by plants into sugars by photosynthesis, and this food is used for growth and development. The plants in turn serve as food for the herbivores, who are eaten by carnivores, and this places plants at the base of the food chain. Photosynthesis is the finely evolved process that makes all life on Earth possible, by converting light from the sun into chemical energy, which then becomes biomass and habitat, abundance and biodiversity. All that we were measuring today in this plot.

Carbon is thus an essential element of life in forests. It is sequestered and stored in the trees and plants, and transformed, transmuted, and stored in the soils and animals. Indeed, about half of the biomass of a tree is carbon alone, and when we measure the size of a tree, we can easily document how much carbon is contained in its great body. Scientists have estimated how much carbon is stored in forests globally—including the trees, plants, and soils—and what the size of these pools is relative to those in the atmosphere and oceans. When we log a forest, we can easily count how much carbon remains behind in the ecosystem, and how much is transferred to the atmosphere, into streams, or into forest products. We can measure the impact of our actions on the carbon pools, and how quickly it takes a forest to recover its stocks.

Sitting down in the shade, I picked up a fallen Douglas fir branch, the soft needles coated with a milky wax protecting them against water loss in this arid climate. *I could use some of that,* I thought. I pulled my cap lower over my sweaty brow. The glaucous sheath had turned the needles to sage green, the light wavelength reflected by chlorophyll. Chlorophyll is the compound—a pigmented protein—that makes photosynthesis possible. It's composed of a brilliant assortment of carbon, oxygen, hydrogen, nitrogen, and magnesium molecules, in a structure similar to human blood. The magnesium in the chlorophyll reflects green light, just as the iron in hemoglobin makes blood run red. Chlorophyll captures sunlight, just as blood captures oxygen. I

Hannah Sachs, eighteen, taking notes along a coarse woody debris transect at the University of Northern British Columbia John Prince Research Forest in Dakelh and Sekani territory, 2017. The thirty-meter tape is stretched across the plot, and her crewmate is shouting data describing each piece of wood that intersects the tape to determine carbon stocks in the dead wood pool. The data is also used to describe fuel load and risk of wildfire. Hannah is wearing long sleeves because the mosquitoes were so thick her arms and face would swell from the bites.

Nava Sachs, sixteen, identifying plant species and estimating their cover at the John Prince Research Forest in Dakelh and Sekani territory, 2017. In some plots, there were over fifty species of trees, plants, mosses, lichens, and liverworts. Nava would eventually become a crackerjack botanist who could tell the difference between a blue, black, and Alaskan huckleberry by the patterns of their leaf margins.

rubbed some needles on my flushed cheeks, inhaling the tart essence, and was immediately refreshed.

I reflected on the way trees and humans are intimately interdependent, breathing in and out each other's essence in rhyme. The trees (the autotrophs, the producers) *breathe in* the carbon dioxide that we (the heterotrophs, the consumers) *breathe out.* In turn, trees *breathe out* pure oxygen that humans *breathe in.* Photosynthesis and respiration, performed by plants, algae, and cyanobacteria, are the foundation of life. These processes provide the food and energy for the rest of the food chain: insects, eagles, wolves, and humans. The products of photosynthesis—carbon, water, and oxygen—are stored in global pools including plants, soils, the oceans, and the atmosphere.

Hannah and Nava came and flopped down next to me, and Jean joined us to take a break in the shade of the old trees. Nava took a swig of water as she settled among the flowers, then passed the jug around. The understory was rich with dozens of species, some common, others that thrive particularly well under closed forest canopies. My daughters had been seeing and touching these plants since they were babies. I brushed my hand against a Saskatoon berry, also known as serviceberry or juneberry. The shrub was taller than me, with blue-sage serrated leaves and white papery flowers that turn to deep purple berries by midsummer. The plants had flowered around my house in April, the same month Hannah was born. By the time she could hold her head up a few months later, the Saskatoons had fruited in abundance, and my beautiful baby girl reached in delight for their juicy indigo berries.

Not far away was a specimen of *Rosa gymnocarpa,* the baldhip rose, which loves shade and blooms pink and lavender flowers that give way to brilliant orange-scarlet hips that provide food for birds and other small forest creatures.

"You can make tea from the stems and leaves," Hannah said. She remembered making it in a pail in the forest with her cousin Kelly Rose when she was little.

I spotted some *Shepherdia canadensis,* or soopolallie, its leathery opposite leaves lit with bright-red bitter berries, beloved by humans and bears alike. In late summer, a single grizzly will eat thousands of these berries a day. I'd sat around a fire with an elder from the Xatśūll

Nation who'd whipped them together with huckleberries and water to make a foamy dessert called Indian ice cream, or *sxusem*. Jean told us that soopolallie could also be used as an astringent by rubbing the berry juice over one's face.

We were sitting on a mound of *Hylocomium splendens,* or step moss, also known by the fairy-tale name of glittering woodmoss. The hardy souls who live off-grid in log cabins in northernmost Canada used to use step moss—and maybe still do—to fill the gaps between logs and keep the bitter cold at bay. Nava counted the number of "steps" in the moss nearest her, telling its age in years.

When we had finished our break, my daughters retraced the transects once more, this time to measure all of the small wood, the branches and twigs, lying on the ground that intersected the crisscrossed tapes. I was glad the girls were taking care of these measurements because it required constant bending over, and my back was sore. I could hear the twigs crunching under the girls' boots as they worked, and I started to worry about fire danger. The flammable little pieces of wood on the ground are called fine fuels, and this kind of heat could whip up an intense ground fire in a thunderstorm.

"Anyone who walks by will think we're crazy measuring all these little bits of wood," Nava remarked.

"I think most of these trees died from root disease," Hannah noted, peeling away the bark from a Douglas fir log. Hannah and Nava already had a keen understanding of the different forces that killed trees, from fungi to insects, windthrow to fire. My mum would be happy that her granddaughters knew enough to survive a few days eating berries and roots in case they got lost, like she did as a kid growing up in the woods. She'd learned from her mum, Granny Winnie, how to snare a rabbit, catch a fish, and pick berries at their homestead near Edgewood.

Jean cut into the base of a snag with her axe and whistled when she exposed the creamy white mycelial fans of Armillaria root disease, whose fruiting body is known as the honey mushroom. The fungus of *Armillaria ostoyae* lives most of its life cycle in the ground, but its hyphae can spread into living tree roots and the root collar, where it could gradually rot the sapwood and cut off sap flow, killing the tree.

A red-backed vole, no bigger than a mouse, scurried along a log to its nest as the girls approached with their measuring instruments. The log was coated in plants, mosses, and mushrooms that had established in the decay. Saprotrophic fungi—fungal species that feed on dead and decaying matter—eventually transfer much of the woody carbon in these logs back into the soil, and this process continues for decades, slowly increasing carbon content of the soil as the forest ages. But how much carbon in soil actually comes from this downed timber? What happens to the wood when the forests are run over with skidders and the trees cut? Will we be able to save more of these important habitats if we leave some of the old live trees behind? I silently pondered these questions as I worked.

AN HOUR LATER, Jean and I had finished measuring the last of the big trees in the plot. They ranged in age from 203 to 253 years old, all having established during a cool wet climatic period as far back as 1763, the year the Treaty of Paris was signed, when Canada became a colony of British rule. By the time the eldest tree had germinated, the Hudson's Bay Company and Christian missionaries had been firmly established in the Indigenous territories of Turtle Island for over a century. Thick-barked and deep-rooted, these trees had been witness to the entire history of colonization, from the dispossession of the people from their land to the outlawing of Indigenous burning and the establishment of the residential school system that took their children. Now the trunks of these trees were tight-ringed, and the knots of fallen branches were hidden beneath the bark, but they still told the stories of this great change in human relationships with the land.

Most of the trees in this forest were Douglas fir, but there were a couple of large interior spruces along the stream and clutches of trembling aspen in the gaps. In forestry parlance, the tallest fraction of trees in a forest are known as dominants and the main canopy trees are called co-dominants. Having earned degrees in forestry and registered as professional foresters, Jean and I were very familiar with the terminology that pervades the timber industry.

I turned to assessing the intermediate trees and saplings that had grown up in the gaps between the giants over the decades.

"There's scads of little trees in this plot," I remarked as Jean and I got to work measuring the heights and diameters of dozens of saplings.

Since Indigenous burning was banned by settlers in the late 1800s and colonial fire suppression efforts were expanded for timber profits, the understory had been filling in with scrubby Douglas firs. Understory trees that would have been burned by ground fires lit in the cool wet spring a century ago were persisting and growing as slow as molasses. Some were almost a hundred years old yet only a few meters tall. Scientists had determined that this undergrowth was a fire hazard and had recently learned it was also competing with the older trees for soil moisture.

I wondered aloud how much carbon was in this throng of understory trees compared to those in the overstory. As usual, Jean couldn't resist a mathematical challenge. She pulled out her phone and did some quick calculations. After tapping away for a few minutes, she discovered that the handful of mother trees in the plot had captured 62 percent of the carbon in the forest, and the dozens of little understory trees only 1 percent. This made sense, as the foliage in the colossal crowns of the mother trees intercepted most of the sunlight and therefore were photosynthesizing at vastly greater rates than the shaded sprites in the understory.

"Listen to this!" Jean walked over and patted the chunky bark of the biggest mother tree. "She contains one hundred and twenty-five times as much carbon as this little tree." She reached over to touch the smooth bark of a nearby understory tree.

My studies had shown that the biggest, oldest trees are the hubs of the forest, connected to their neighbors through a vast network of mycorrhizal fungi. These fungi are crucial to the fitness of the trees by providing them with nutrients and water. In turn, the fungi depended on the trees for photosynthate, enabling them to grow their vast threads through the soil pores in search of nutrition. The fungi could also link the trees together, providing them with multiple sources of photosynthetic energy. The big old trees are more densely linked than the others because of their massive root systems with innumerable points of connection. I had reasoned that their enormous crowns are also the energy centers of the forest, carrying out the vast majority of photosynthesis and redistributing some of this energy to smaller trees and saplings through the mycelium.

The last plants appraised for the day, we began to wind up our tapes and gather our gear. Jean unfolded the creased map to plot our way out. Even Hannah and Nava were feeling the late-afternoon heat as they doused their heads with water and expressed relief that we were done. The sky began to grow dark as a thundercloud swept in overhead. Lightning flashed in the distance. Seconds later, we heard a rumble like a logging truck. I glanced nervously at the mountain ridge above. The trees stood motionless. Jean raised her head to sniff the air.

I looked to the east and saw a plume of smoke rising over the ridge. *Oh no.*

"Time to get out of here!" I said.

We quickly stuffed the increment cores and hand trowels into our vests. Within minutes, the gathering easterly wind was bringing smoke into the stand.

Hannah hastily filled in the soil pit. Ghostly tendrils of smoke were already drifting around the trees like scarves unfurling.

"Just leave it," I said to Hannah as she tried to scrape the last minerals into the hole. "Let's go!"

Jean struggled with her balance, so Nava snatched Jean's pack and I took her vest. Hannah grabbed the sample bags. A deer ran through the plot. Then a rabbit, bounding at top speed. We grabbed the rest of our gear as the smoke started to thicken.

"Run!" I shouted. "Don't look back. Keep moving!"

Ahead of us, Hannah and Nava sped like deer, leaping over logs, jumping the parched stream beds. I held Jean's arm tightly. Together we ran through the aspens and thickets of fir, shouting at the girls to keep going, we'd catch up at the truck, still half a kilometer away.

I glanced back to see the growing fire licking through herbs and shrubs and consuming fine woody debris on the forest floor. I could see it igniting the dead branches low down. The blaze moved swiftly, hissing and crackling, hungry for fuel. The fire quickly climbed the interwoven ladder of suppressed trees into the intermediate trees. In an instant, it had jumped into the resinous midstory of the canopy trees.

Jean tripped, and I helped her up. I could see fear in her eyes.

"C'mon Jean, we can do it."

I could hear my pulse hammering in my ears as I ran, and I started to feel a stitch in my right side, stabbing me in the ribs with each breath. I snuck another glance back: the fire was now moving through our plot. The wind had picked up, and flames were racing into the canopy. With a whirling eddy of oxygen-rich air, the fire ignited the tree crowns, burning into the living wood.

The big tree we'd just measured candled. The fire in her crown spread to the neighboring tree, and the next. The fire was on the move. With the high temperatures and dry wood, the forest was turning into a curtain of flames. The center pin of our plot would survive no matter how destructive the fire. It was now a fire-severity pin. It would tell this story later to anyone willing to listen.

A gray wolf emerged from the trees, running in the same direction we were. He loped alongside us, his tongue cooling him as he drank in oxygen. He squinted at me through the thickening smoke and I could see his amber eyes.

With the flames chasing us, we charged through the trees and burst through the timber edge where the girls were waiting anxiously at the truck.

"Mum!" Nava cried. Hannah grabbed our gear and threw it into the truck bed. I wanted to gather my daughters up in my arms, but there wasn't a minute to spare. Jean flung open the door and slid into the driver's seat, turning the key at the same time, and the engine roared to life. She wheeled the truck around and I radioed to report the fire to the dispatcher. As we drove away, we saw the wolf disappear into the trees.

We reached the highway half an hour later and stopped to collect ourselves. Helicopters were hurrying toward the tower of rolling smoke rising from the woods, and we watched the choppers scoop buckets from the lake and fly toward the flames. The fire seemed uncertain of what to do next. Then its flames twisted into spirals and gyres and into a single tornado, painting the sky a glowering shade of orangish black.

We turned onto the highway, and knowing we were safe, Hannah and Nava began to chatter with excitement about our close call. I twisted around in the front seat and examined my daughters. Both girls were enveloped in a cloud of soot and grime. Nava's T-shirt was so caked with dirt it was impossible to tell what color it had originally been. Soil was ground into Hannah's knees where she'd taken a spill as she ran.

Jean smiled. "You guys look like Pigpen."

"Look in the mirror!" Hannah cracked. I tilted the rearview mirror to take a quick look at myself. Face covered with ash, hair matted with sweat, the Creature from the Black Lagoon stared back at me. I shrugged and glanced at Jean, who looked like a mirror image of me.

A long line of cars was accumulating along the shoulder as people stopped to watch the fire. We headed north along the 97, passing through the familiar flaxen grassland, the river rushing along its willowy spine, the tender forest meeting the meadow's edge as the land rose into the mountains. My mind wandered to my brother, Kelly, herding cattle in this valley a couple of decades ago.

"Want to visit?" Jean whispered. She always knew my thoughts before I had to say a word.

I nodded, and she drove a few more kilometers to the north, turning onto the dirt road that led to the Onward Ranch. We crossed the stream and parked at the cattle gate that marked the path leading to Kelly's knoll near Yellow Lake.

"You go on ahead," she said. She pulled out her binoculars to watch the choppers now circling like dragonflies. The girls were born after Kelly died, and they'd only been to his knoll once, but they knew their uncle through stories. They stayed with Jean to watch the ballooning smoke from the fire. I told them I wouldn't be long and started down the path.

When I reached Kelly's stone, I brushed away the fescues to reveal the baldhip rose engraved on the copper plaque. I leaned against the old Douglas fir mother tree bent over from years in the wind, her bark calloused from the ground fires that had swept through on occasion. She sifted her evergreen breeze over my shoulders. I missed my brother, though he'd been gone twenty years. I gazed over the rolling landscape, at the hay-filled valley, ochre, yellow, and brown. The whitewashed St. Joseph's Mission, the now-shuttered Indian residential school. Kelly's Indigenous friends had told him about the atrocities committed at the school by Catholic priests, and I remembered how it had pained him to work at the site of a place that had witnessed so much heartbreak and loss. Years later, the unmarked graves of at least one hundred and fifty-nine Indigenous children would be found on the school grounds. It was there, at the old gymnasium beside the mission, where Kelly had breathed out his last words in the accident. He'd been run down by his own tractor—it had slipped into gear while he stood in front talking with some kids—sending his spirit to join with the mountains he'd loved.

The Douglas firs on the slopes were now thirsting from drought, weakened by fire suppression, spruce budworms, and bark beetles. They were missing the people who had once cared for them. Kelly and I were surrounded by 19 million hectares (47 million acres) of dead or dying forest—a boneyard as big as France.

I was glad my brother wasn't here to see these forests suffer. Or the decline of forests worldwide, struggling to survive in conditions they had not evolved to cope with. Canada was warming at three times the global average. The fires had become more intense, more extensive, more severe. In the past three years, over 3 million hectares (7.4 million acres) had burned in British Columbia alone, the fires stoked by fuel buildup caused by decades of fire suppression and the hot, dry weather that had become the new normal.

I picked a single stem of *Rosa gymnocarpa* from beneath the old fir and placed it on Kelly's gravestone. Our ancestors had lived on the land all their lives, and Kelly had, too. We had tried to tread softly and tend the Earth with respect, working alongside the First Peoples who were there long before us. I whispered a promise to continue this work, to live in right relation, to help make the lives of the next generations easier. Would protecting, restoring, and tending forests help restore the balance so badly needed in our world? I looked across the valley at the stressed landscape and I resolved to find out.

2

KINSHIP

I pulled open the double doors of the UBC greenhouse and was immediately enveloped in its warm, humid embrace. This early in the morning, the earthy old building was nearly empty except for a student sorting seeds at a bench in the corner. I could hear the soil-sifting machine—a truck-sized box of fabricated steel that worked like a cement mixer—churning away in the dark, damp side chamber. The machine rattled to a halt and the chamber door opened. Amanda appeared, wearing a white lab coat dappled with soil crumbs and hoisting a tub of soil. I hurried over to help her with the heavy container, and the two of us wrestled it to the table where she was potting up her experiment.

I always looked forward to spending time with Amanda, who had become my protégé and friend over the past decade. I'd had many brilliant graduate students, but Amanda stood out. Her sharp intellect was balanced by the warmth of her personality, and she got along with anyone and everyone. But it was her sense of humor I loved the most, her playful wit and easy smiles and laughter. We always spent the first five minutes of her visits to my office cracking jokes.

"Only ten hours planting the pots?" I'd tease, knowing she'd been up all night establishing her latest experiment.

"Just me and the mice," she'd say, having made friends with the nightly parade of creatures in the greenhouse.

As busy as she was, Amanda always stepped up at the first sign of need. "What can I do to help?" were typically the first words out of her mouth. She spent countless hours teaching new graduate students how to sort roots and count mycorrhizas, and she burned the midnight oil as a teaching assistant marking hundreds of undergraduate exams.

"How do you think these seedlings are doing?" she asked me now, leading me through the sliding glass doors of the first greenhouse bay. "I'm worried some of them have a disease."

I looked at Amanda's bench—twenty meters long and two meters wide—which was covered with hundreds of the four-liter pots in which she'd planted groups of Douglas fir seedlings. Each pot was like a small forest where she could investigate the relationships among a community of trees, some strangers and some kin. The pots were carefully labeled with intricate codes keeping minute track of each treatment. Frankly, I was floored by the complexity of Amanda's experiments. Tracking each seed, seedling, neighbor, mycorrhizal network treatment, and pot required a highly focused and precise mind. I examined the Douglas fir germinants she was pointing to. I could tell by the leaf discoloration that some of the seedlings had a wilt infection, but there was little mortality, so I told her I wasn't too concerned.

"I don't think I have enough of Family A coming up in this pot," she said. "Should I move these over?"

Using tweezers, I helped Amanda transplant the tiny germinants from one pot to another. I enjoyed working in the greenhouse. Transplanting seedlings—putting my hands in the cool soil, breathing in the sweet earthy fragrances—was soothing and meditative.

"How was the World Cup?" I asked.

"We won the silver!" Amanda described the lightning strike that had hit the outfield in the last inning, but didn't mention she had hit the two-run double that resulted in her team's medal.

"But how are you? How have you been?" she asked, changing the subject. I knew she wouldn't offer up more details unless I pressed. Apart from her brilliant forestry research, Amanda was a star in the sports world. She'd played NCAA ice hockey and softball at Brown University, and CIS hockey with the UBC women's team, the Thunderbirds, but baseball was her true love. She was the longest-serving member of Canada's women's baseball team, which was ranked second in the world. That year, 2017, she earned a spot on Baseball America's list of the ten best female baseball players in the world, ranking seventh and the only Canadian. Amanda had quietly been winning international championships and medals for years, though she never said much to her colleagues at the university about these incredible accomplishments. Mostly she just wanted to talk about the enormous datasets she'd been analyzing, her spreadsheets as detailed as the stock exchange, keeping track of thousands of individual seedlings in her kinship experiments.

The topic of Amanda's research—kin recognition—was fitting. She was a teammate first and foremost, and she had applied that lens to her studies. Her natural inclination as a team player had led her to question how the community of trees in the forest influenced kin relationships. Her doctoral work looked deeply into what the composition of the forest meant to relationships between kindred Douglas fir seedlings. Might a tree's neighbors—the species, density, or even the soil they grew in—affect how well Douglas fir relates to its own kin?

This research built on her master's discovery, where she'd found that Douglas fir distinguished between neighbors that were related—kin—and those that were strangers. Douglas fir seedlings growing in the presence of relatives were healthier and more vigorous and had greater mycorrhizal colonization, and seedlings received greater carbon subsidies to their mycorrhizal root tips from nearby kin than from strangers. These findings hinted that Douglas firs could enhance the growing conditions of relatives over outsiders, perhaps making "elbow room" by changing the competitiveness of their roots, or improving conditions in the soil to increase the chances kin neighbors would survive. They could help explain why Douglas fir seedlings grew clustered around parents in the dry interior forests of British Columbia. The facilitation of kin could also enrich our understanding of why some tree species grew in relatively pure stands in temperate forests, distinct from the vast diversity of trees that grow together in tropical forests.

I plunged my tweezers into the soil and gently pulled up a seedling, careful not to disturb its tiny taproot as I transplanted it into a new pot.

"So, uh, Amanda—how's that paper coming?"

"Ha!" With a wink, she assured me she'd get it done after the

Pan Am Games. This was a long-standing refrain between us, started when one of her committee members kept asking when Amanda was going to publish. Doctoral students were under tremendous pressure to publish in peer-reviewed journals. It was just like Amanda to use humor to defuse this nerve-wracking rite of passage. Underneath the playful banter was a firm resolve: Amanda would not be rushed into publishing the results of her research prematurely.

We heard the greenhouse doors swing open and my master's student Eva joined us, lugging a green satchel crammed with her laptop and textbooks. Eva had started her university studies late, at the age of twenty-six. She had taken my Intro to Forest Ecology class as an undergrad and approached me after class one day. "I love the way you describe ecology," she had said, fixing me with her serious blue-gray

Amanda Asay (right), twenty-two, and I had just finished measuring seedlings at an experiment investigating how canopy gap size affected interior Douglas fir regeneration in Tk'emlúps te Secwépemc territory, 2011. Amanda and I shared a goofy sense of humor. "Play it by ear" was one of our favorite sayings, as we innovated and adapted during her groundbreaking research on kin selection in Douglas fir.

eyes behind the round lenses of her glasses. She wanted to know if I needed help with anything, and since I had just gotten some funding to start a new project, I hired her on the spot. Eva would become the first of dozens of students to work on the Mother Tree Project in the years to come, spending the whole of that first summer working out of a small cabin in the Cariboo Mountains.

"Hey Eva!" Amanda held up her hand for a fist bump. I listened as the two young women compared notes on their field studies. Their commitment and camaraderie reminded me so much of me and Jean in our twenties. Though Eva and Amanda were still getting to know each other, I hoped the spark of friendship would catch.

BACK HOME IN NELSON, I decided to drive out to Kootenay Lake—one of the eight experimental forests we'd established so far—to catch up with the forester supervising the logging show.

As Jean and I worked on selecting our study sites, we had identified the logging companies working in the areas and asked if our study could piggyback onto their plans. Would they be willing to adjust their logging to fit the parameters of our experiment? Most had readily agreed—especially at the Kootenay Lake site, whose tributaries were water sources for people living in the valley, so the logging in this area had been contentious. Participating in the Mother Tree Project would help the company's public image: they weren't just taking trees, they were also helping care for the forest. However, the reality was a little trickier. I had negotiated the logging of our sites with the supervisor, but the logging foreman, the guy on the ground, was the one actually carrying out the orders. Most of these fellows were lifelong loggers and clearcutting was what they'd always known. Leaving patches of big, valuable Douglas firs uncut—as some of our study treatments specified—went against what they were used to. But it was crucial that the loggers followed our prescriptions to the letter or they wouldn't be comparable across the whole range of our research forests.

Kootenay Lake's moderate climate landed it smack in the middle of our climate rainbow, which ranged from hot and dry in the south to cooler and wetter farther north. Climate is the largest-scale influence on the regeneration of forests and the primary driver of natural selection that shapes local adaptation of trees. Climate variation is the factor that ensures whether a tree can respond and adapt to the environmental ups and downs of its home base. The natural climatic variation among our experimental forests would give us insight into how the increase in aridity with climate change might affect how well our forests coped. Observing how trees grew in different regional climates was a much faster approach to predicting climate change effects than watching our forests gradually become more arid on their own. Lower down on the hierarchy of influences, the local forest conditions themselves—the overstory trees, the soils, and the lay of the land—were the next drivers of forest regeneration.

Also critical to our analysis was the site's history of disturbance. Had a fire swept through the forest, and if so, how long ago? What patterns of trees remained after the burn and did they help the forest regenerate? Or had the forest been logged, and how recently? Could harvesting practices be designed to emulate the natural disturbance patterns so the legacies—old trees, plants, seed banks in the forest floor—aid in the recovery? Jean and I, with the help of local experts, had designed our five harvesting treatments to help us determine how forests respond to and recover from disturbances like clearcutting or partial cutting in a changing climate, and what tree retention patterns would result in the most successful forest regeneration. We hoped our treatments would tell us how many of the mother trees needed to be left in each climatic region to help protect the integrity of the forest, and what spatial patterns of retention were most favorable for buffering both naturally regenerated and human-planted trees from the weather.

At one end of the tree retention gradient, we clearcut all of the trees—in other words, leaving no trees behind, zero overstory retention. Clearcutting was the standard operating practice by the forest industry in the province, so it needed to be represented in our experiment. Just as importantly, it would create the most extreme environmental conditions against which to compare our gradient of overstory retention treatments and uncut control forests.

In the next treatment, we retained 10 percent, or about twenty-five of the largest trees per hectare, with the remnant trees evenly dispersed about twenty meters apart. These scattered mother trees would pro-

The clearcut treatment, aerial view, in T'exelc te Secwépemc territory at the Alex Fraser Research Forest. Clearcutting is where all trees are cut in an area at least two hectares in size, with at least half the area fully open to the sun. The clearcut treatment plots in the Mother Tree Project are four to five hectares. They are small in comparison to most clearcuts in British Columbia, which are typically ten to thirty hectares and sometimes reach a thousand hectares (2,500 acres) in salvage-logged, beetle-killed, or burned forest.

Seed tree treatment, aerial view, in T'exelc te Secwépemc territory at the Alex Fraser Research Forest. Approximately twenty-five of the largest Douglas fir trees per hectare, or about 10 percent, have been left standing, and all else harvested. The remaining trees disperse seed that facilitates natural regeneration.

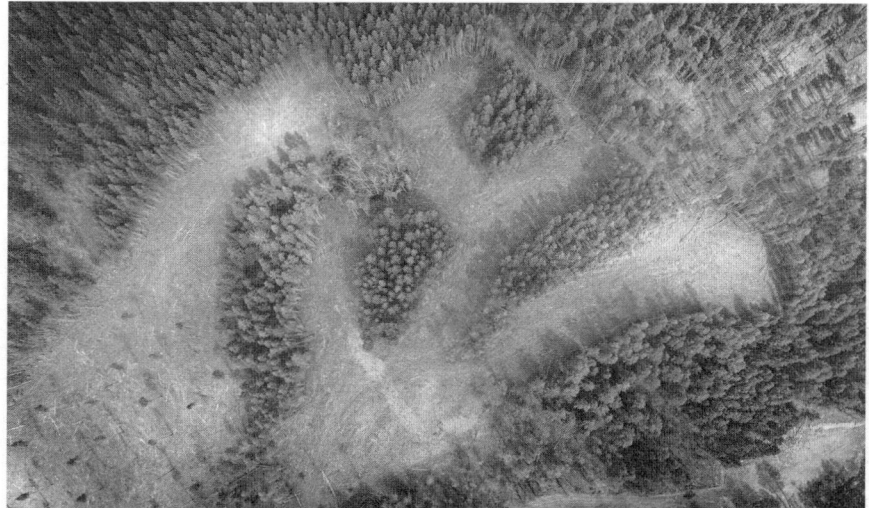

Thirty percent retention treatment, aerial view, in T'exelc te Secwépemc territory at the Alex Fraser Research Forest. Patches of forest about thirty meters across have been logged using feller bunchers and grapple skidders, leaving clusters of intact forest approximately thirty meters wide. The retained patches are centered on the biggest Douglas fir trees and their immediate neighbors.

Sixty percent retention treatment, aerial view, in T'exelc te Secwépemc territory at the Alex Fraser Research Forest. The feller bunchers cut the mature trees in trails through the forest and reached in with the long-armed saw to cut smaller trees. This method leaves a matrix of about 60 percent of the largest trees and nearby canopy neighbors, while creating small gaps for natural regeneration.

vide seed for the next generation in a technique commonly known as the seed tree method. The mother trees would also serve as hubs for mycorrhizal networks that the new seedlings could tap into for fungal inoculation. The seed trees would harbor the diversity of organisms in the soil, promoting resilience of the new regeneration against stresses, and ensuring at least some species of fungi remained that were compatible with any new seedling recruits.

In the 30 percent retention treatment, we protected islands of mother trees about thirty meters across, while clearing away small patches of surrounding trees in a system called aggregated retention. The harvested forest looked like Swiss cheese, with the islands of intact trees being the holes in the cheese. The islands contained a few large mother trees and their immediate neighbors. These clusters of trees would provide seed for the next generations, and many of their offspring could establish under the influence of their parents. The neighborhood trees would also protect one another against the elements compared with the seed tree treatment, where the retained trees stood alone, and the open gaps would provide ample sunlight for the new recruits.

In the 60 percent retention treatment, we retained the tallest canopy trees while selectively harvesting a few mature neighbors and thinning out the smaller trees below. The retained trees formed an integrated irregular canopy, providing shade, seed, mycorrhizal fungal inoculum, and plenty of shelter for plants and animals. The newly establishing seedlings would reflect the species diversity of the retained canopy, and would also include a range of Douglas fir kin and non-kin genotypes. This technique was similar to the continuous cover forestry often practiced in Europe. We thought it might also mimic the results of a low-intensity cultural burn conducted in springtime, when fire would be used to clear out the understory while leaving the big trees standing.

Finally, in the fifth treatment, we left the forest untouched as a control, with all of the trees still standing. As with the clearcut treatment, the uncut control represented the extreme end of our range of treatments, but at the opposite end of the retention gradient. Leaving the forest intact would allow us to learn how the forest developed without human intervention. We repeated these five harvesting treatments up to four times at each of our experimental forests, to better confirm our findings.

So far, only one replicate of our five treatments had been harvested at this site, and this was my chance to ask for some improvements.

I navigated the winding road carefully, passing the glistening lake and yellowing cottonwoods. At the start of the logging road, I radioed my location on the two-way to signal any oncoming trucks. I motored slowly around the curves, meeting small patches of snow in the shady stretches, the wide ruts telling me the road was active. A logging truck called its location a kilometer uphill, confirming my hunch, so I squeezed to the side as the massive truck roared past me, a half dozen old-growth cedar logs shedding bark and branches in its wake.

Watching the truck and its precious cargo disappear in my rearview mirror, I suppressed a shudder. I knew it had been the plan of the colonial government over the past century to cash in on the primeval forests of British Columbia. But the pace of harvesting was adding greenhouse gases to our atmosphere and destroying the habitat that bears, caribou, and owls needed to survive. It was magnifying fire risk and increasing the frequency of floods. Our planet's climate was changing, and this old operation that was part of my trade seemed to be amplifying the problems.

I sighed, reflecting on the history of logging in British Columbia. The rapid clearcutting of Canada's forests was well underway when I was growing up in the 1970s. The first clearcut appeared above Mabel Lake, where my family lived, when I was in my teens. I was shocked to see trees that had taken hundreds of years to grow felled as rapidly as day turns to night. The path had been paved three hundred years earlier in 1670, when King Charles II of England licensed the Hudson's Bay Company to seize the resources from the Indigenous people. As with the fur trade, the forests were exploited to feed European wood demand and foster economic development of the created colonies that would become the nation of Canada in 1867.

The government began regulating forest harvests in eastern Canada in the 1820s. By 1949, the chief foresters of the provinces determined the allowable annual cut rate for their jurisdictions—under advisement of the private companies granted licenses to harvest trees. The allowable cut rate was the amount of forest legally available for cutting, which was intended to balance environmental, economic, and social considerations, and theoretically match the growth rate of the forest. No more was to be taken, and no less, to ensure a long-run sustained yield, unless growing circumstances changed. If climate changed and became less favorable for growth, in theory the government would reduce the cut to match the new conditions. However, in practice, this kind of reduction was rare, as mills and jobs created to service the cut depended on maintaining historic cut rates.

The vast forest and cool climatic period between 1947 and 1976 at first buffered the high rate of cut against larger ecosystem level changes. But this changed in the 1990s, when surface temperature flipped to a warm phase in western Canada and climate change accelerated. The mountain pine beetle population had exploded during an exceptionally warm winter in 1991 and the insects chewed through an astronomical 18 million hectares (44.5 million acres) of forest over the following decade.[1] Scientists established that the unprecedented outbreak was triggered by the warming climate, which sped up beetle reproduction, and a continuous policy of fire suppression, which enabled the spread of pine trees across a vast area during the colonial settlement period.

The chief foresters did not reduce the rate of cut to account for this vast loss of forest to the beetle. Instead, the cut was "uplifted" to clear the way for salvaging the dead, dying, and, through convenience, surrounding healthy trees. The timber companies reaped the profits of this uplift, which was implemented to salvage the damaged wood before it lost its commercial value. The cumulative mega-disturbance, a combination of human-caused climate change and extensive salvage logging, pushed the landscape into a degraded state. A degraded forest is one that has been so damaged by human activities like excessive logging, road construction, or wildfires that its ecosystem services, like carbon storage and biodiversity, are disrupted.[2] No longer was there sufficient primary forest to buffer against the "sustained yield" harvest rate.

By the early 2000s, increasing drought from climate change diminished forest productivity at the national level. Along with this came outbreaks of spruce-bark beetle, Douglas-fir bark beetle, and western spruce budworm. The warming temperatures had created a more favorable environment for bark beetles, speeding up their reproduction and expanding their range. Normally, a healthy tree can "pitch out" attacking insect pests by excreting a sticky resin, but the drought-weakened and heat-stressed trees were unable to mount much of a resistance against the invading hordes of beetles, and trees perished by the millions.

In 2001, as a result of growing drought and disturbance, the forests of Canada shifted from being net sinks of greenhouse gases to net sources. With logging targeting damaged stands, existing clearcuts merged into even bigger clearcuts, and the spread of even-aged plantations made the landscape even more vulnerable to catastrophic fire and pests. Precaution as a guiding principle, however, was ignored, and the clearcutting continued apace, even in healthy forests.

These industrialized forests, combined with hot and dry conditions resulting from climate change, contributed to the megafires starting in 2017, and in some areas, the fires were followed by flash floods.[3] The steeper soils, less able to absorb and retain water, in some cases quickly eroded and flooded during spring freshets, and then dried out later in the season with summer moisture deficits. These massive disturbances, along with heat waves and drought, were now causing many tree species, particularly in mountainous regions and in the uplands just below the treeline, to decline—not just in Canada, but across the Americas and in Europe.

I knew that the century of timber exploitation was a significant contributor to the climate crisis stressing our forests. I was deeply worried. How much forest carbon was being lost and emitted with clearcutting and extreme forest fires? How were these greenhouse gases contributing to climate change and in turn further weakening our forests? I hoped the Mother Tree Project would illuminate a better way forward.

As I got closer to our experimental site, my ears were filled with the buzzing of chainsaws and roar of heavy machinery. A skidder bounced down the spur road pulling a bundle of cut trees, dragging the crowns along the berms and ruts, leaving branches and needles in their wake. The air was sharp with the scent of resin as I made my way through the slash to the landing.

Hank, the logging foreman, walked down the skid road to greet me.

"You did a great job with the clearcut treatment, Hank." I pointed to the five-hectare opening where all of the trees had been cut. Privately, I wasn't keen on this treatment, but I was glad the clearcut was small and I knew it was crucial to include clearcuts in our experiment for the purposes of comparison.

"Thanks," he said, taking his cap off and scratching his patch of short gray hair.

"And I like how you did the sixty percent retention treatment." The fallers had cut strips a skidder width through the forest, leaving the stretches of forest between the skid roads, about two tree lengths wide, intact.

"Yeah, but it was hard reaching into the patches to thin from below." Hank explained how the arm of the feller buncher—the motorized harvester that cut and gathered trees—could only reach five meters into the forest to cut the understory trees.

"It's good for the forest, though," I said. Leaving 60 percent of the trees and limiting the skidders to designated trails would protect the carbon pools in the trees and the soil, and leave the mosses and lichens and forest-dependent herbs intact.

I drew a graph in my notebook and showed him our predictions that each additional tree removed should reduce the carbon stocks in the aboveground pool incrementally, to where none of the aboveground carbon was left in a clearcut, other than the logging slash. He looked skeptical.

"It's costly for us to do that treatment," he said, pointing to the sixty on my map. "We can't make enough profit with so few trees cut, and it costs us more to design the skid trails."

I reluctantly agreed that, at the end of the project, my research grant, which was barely sufficient to cover our field technicians and sample analyses, could pay for any additional logging costs. I'd have to find the dollars somewhere.

The skidder driver was yelling at Hank, trying to get his attention. He turned to go.

"Uh, one more thing, Hank."

I showed him the two treatments where I'd asked for lower retention levels—in other words, leaving fewer trees standing. One was to leave the twenty-five largest seed trees evenly distributed across the block. The other treatment called for retaining 30 percent of the best trees, leaving patches about one tree length, or thirty meters wide, intact.

"You left the worst trees behind, not the best ones," I pointed out.

I could see his left eye twitch. He laughed. "Yeah, but it kills me to leave these big Dougs."

"You've taken the best Douglas firs and left the spindly ones and the little cedars."

"Well, the big trees will just blow over in the wind," he countered.

I forced myself to take a deep breath before answering. I reminded myself how important it was to cooperate with the loggers.

"I need you to do a better job of these for the next three reps," I said. "Otherwise, I can't compare these treatments to those at the other sites."

Hank pushed back. "We are doing our best, but we still need to make a profit," he said. "Even if the treatments are part of a research project."

The next day the head forester called me at home to say the contractor was moving their equipment to cut a higher elevation forest and would return to finish the logging of my experiment in a couple of months.

"But I've ordered trees for spring! I need you to finish logging the experiment this fall so the ground is ready for us to plant at spring breakup."

If the sites weren't logged, we would not be able to plant.

"Well, you shouldn't have ordered your seedlings so early," he replied.

I hung up in frustration. The snow was already creeping down the mountainsides.

DAVE WALKEM is an old friend from the days when Jean and I worked temp jobs describing forest ecosystems for the government of British Columbia. When we met, I was a twenty-two-year-old summer student while Dave had recently graduated from the Faculty of Forestry at the University of British Columbia and become the first status Indian to be registered as a professional forester in the province. I remembered the research group celebrating his success in the government office, Dave smiling broadly among friends, and Jean and I, both new to the job, wondering if we would ever become professional foresters ourselves. Since I'd last seen him, Dave had served for twenty-eight years as chief of the Cook's Ferry Indian Band, one of fifteen individual bands of the Nlaka'pamux Nation, the Interior Salish peoples. He had earned a master's degree in business, which provided him the background to help his people negotiate with the provincial government to access their own traditional territory and resources. Today, he sat on the band council.

He picked up my call on the second ring, his familiar "Hello" warm and gentle, just as I remembered. After catching up, I told Dave about the Mother Tree Project. I explained that we were designing partial cutting systems we hoped would protect the underground networks connecting the forest, instead of continuing to take all of the trees as the big companies had been doing. That we wanted to learn how to better manage the Douglas fir forests in the province rather than clearcutting them.

"I'm wondering if you'd be interested in hosting one of the experimental forests in your territory and collaborating on the project?" I asked. Nlaka'pamux territory, at the confluence of the Nicola and Thompson Rivers, included some of the most arid Douglas fir forests in the province, and a collaboration could extend the climatic range of our network of experimental sites.

"That's interesting," Dave said. In the silence that followed, I could hear him thinking. I knew that clearcutting was a sensitive topic for him.

Dave's father, as with many in his nation, had been a logger for the hardscrabble independent sawmills that sprang up in their territory in the 1950s. My own ancestors were similarly small-time loggers at Mabel Lake, about 300 kilometers (185 miles) to the east. But in the 1960s, the provincial government failed to renew these small licenses and instead awarded the tenures to big companies, doing away with the small operators and putting many people, Indigenous and non-Indigenous, out of work. Few of the big companies hired Indigenous loggers, and the government did not award any tenures—the right to harvest timber—to First Nations people.

"They were giving away our resources in our backyards and not giving us the opportunity to get our foot in the door," Dave told me.

Thirty years later, in the mid-1990s, after pressure from Indigenous groups, the government started releasing forest tenures to First Nations. But there were conditions.

Dave was negotiating access to his nation's own resources with the government, but they wouldn't award them tenure to cut the forest by themselves.

"They said we weren't good enough managers to look after the forest, so we had to partner with a non-Native company," he said. "They told us that First Nations were too risky to look after any volumes, and so they gave the tenures to the big corporations."

Dave would go on to help set up two joint ventures between local Nlaka'pamux Bands and two nearby sawmills in order to access timber in their own forests. In 1997, the province also set up a pilot project called the Innovative Forest Practices Program in the Merritt Timber Supply Area, which included Nlaka'pamux territory. This program awarded increases in the allowable annual cut for timber companies with replaceable forest licenses where they had implemented innovative forest practices that increased forest growth, and where companies were in agreement with local First Nations.

By 2004, a 50 percent increase in allowable annual cut was awarded under this program, and Dave was able to negotiate access to a small replaceable forest license for his people. He established Stuwix Resources Joint Venture among eight First Nations, and he obtained a forestry co-management agreement among all forest companies, First Nations, and the province of British Columbia. The Stuwix Resources Joint Venture then helped set up six First Nations logging contractors to work for Stuwix and the other licensees.

These contractors quickly became some of the best, most trusted logging contractors in the area. All they'd needed was an opportunity. This was a pivotal time for First Nations, when they were finally getting access to licenses for harvesting timber in their own territories. It was also good for the whole forest industry, and for the forest itself, because it incentivized innovation and cooperation.

Dave explained that this opportunity didn't last long, however.

Focused on short-term gains, the province failed to award the innovation as replaceable licenses. Once the licenses were up, the incentives for improved forestry disappeared, and the forest companies halted their transformative practices. Worse, the First Nations people were left with no replaceable tenure to manage. The big companies quickly reverted to their old ways, resulting in extensive clearcutting of the beetle-damaged pine forests over the ensuing decade. Once the pine forests were gone, the forest companies continued the clearcutting practice in the lower-elevation, uneven aged Douglas fir forests in Nlaka'pamux territory. For Dave's people, this was a disaster.

"The Douglas fir forests are our breadbasket," he told me. "They are a big chunk of our territory. That's where a lot of our medicines, our food, the animals we harvest live, where we are tied to the land."

He paused.

"Can you come to Spences Bridge so we can talk?" he said. "I gotta run now, but I'm interested."

THE NEXT MONTH, Teresa, Jean, and I met in Spences Bridge, a tiny mountain community in the interior of the province between Lytton and Cache Creek. We stayed overnight at the old Gold Rush–era Inn at Spences Bridge, and that evening sat on the veranda overlooking the Thompson River, its glacial current moving so swiftly that one eddy emptied into another like pots of boiling water. As we sat in the cool breeze, we watched a bald eagle dip into the river, then rise swiftly above the sharp sage arêtes, a rainbow trout in its talons.

I loved this dramatic landscape and had conducted many trials and experiments in these forests during my time as a research scientist for the Ministry of Forests. I had one site up Murray Creek, a tributary to the northwest, whose canyon rose steeply into Douglas fir forests, and higher yet into the montane spruce and alpine meadows. That experiment examined how to establish lodgepole pine seedlings in the thick pinegrass that flourished after the dry Douglas fir forests were clearcut. I had no control over the cutting method in that experiment, so I was trying ways to overcome the harsh limitations imposed by the removal of all the old trees. Now I hoped to find a way to keep the mother trees so they could naturally rejuvenate the forest on their own after some partial cutting that selectively removed the smaller trees.

The next day, Dave greeted us at the band office.

"When you called me last month, I hadn't heard of the Mother Tree Project," he told us. "I honestly thought it was a pipe dream! But what you said—how the plants and trees are connected through underground networks—it doesn't surprise me that the science has caught up to that."

Teresa nodded to him. "I agree. This is something Indigenous people have known for thousands of years," she said. "This project can allow us to show how these mother trees are connected belowground with each other, with their communities."

She explained that the Mother Tree Project would demonstrate Indigenous knowledge. We would get to see how the complexity of the networks belowground, and the species interactions, enabled the trees to connect and communicate.

"In our culture, the Nlaka'pamux, these connections are part of our creation story," Dave told us. "At the beginning, the plants, animals, and people were essentially the same, interchangeable. They could transform back and forth from plant to animal to human and back again. But at some point, the transformers announced who were going to be the people, who the plants, who the animals, and so on. It didn't take away the connections we have with all our relations, but we became more of who we are today."

Teresa interlaced her fingers to demonstrate. "When the roots are connected belowground, it's like our communities; when we are all connected, we are stronger. And likewise, the forest is stronger when the trees are allowed to have their connections with each other."

"We believe, as Nlaka'pamux people, that when you harvest a tree, a deer, or a berry, you talk to the people, our relations," Dave said. "And you thank them for giving us their life, for offering the food, as a way of showing respect. Because we are not all just living in a silo—we are connected to the land around us."

Jean and I listened, nodding. These were values that we, too, understood and aligned with in our forestry work.

"The industrial management of forests in British Columbia does not include First Nations values," Dave pointed out. "The innovative silviculture we want to practice would include management for our values on the land. We would keep the different uses, for food and medicines and spiritual places. We want them acknowledged, valued, and respected. Unfortunately, that has not happened in the past. The way forestry is practiced by the industry is not done properly in our sense.

"Now, we have to go and fight to make sure there isn't clearcutting in the dry Douglas fir areas," he added. "The industry says they don't know how to do any other kind of logging. But I'm sorry, it's not rocket science.

"The forests are not being managed for the local ecosystems or the local people," he continued. "They are being managed based on the values of the big corporate owners. These corporate values do not reflect our local values."

"Perhaps together we could design partial retention methods that protect those values while still making wood available to support livelihoods from the forest," I said.

Jean pulled out a sheaf of paper showing the configurations of our five Mother Tree treatments. All but one were designed to protect the existing plants, animals, and medicines. The retention treatments emulated a mixed fire regime that reflected the historic disturbance pattern in these forests, removing the small trees and leaving the old thick-barked, fire-resistant trees standing. Dave studied the treatments in silence for a few minutes.

"The only thing I would want you to change for our territory is to not include a clearcut treatment," he said. "The old trees regenerate the forest. If you take them all, the medicine plants won't come back."

Jean and I shared a look of assent, and I could see Teresa nodding, too.

"We can leave behind the biggest trees," Jean agreed. We could still create open spaces in smaller areas to test for differences in climatic fluctuations.

With Dave's blessing we added the ninth, most arid forest, located at Twaal Creek, to the Mother Tree Project. In 2020, the First Nation company Stuwix Resources Joint Venture would conduct the harvesting, removing the smaller trees and leaving the old mother trees behind, as Dave had requested.

We were fortunate that Stuwix Resources Joint Venture was still in operation to apply these innovative treatments, even though the province had abandoned the co-management opportunities years earlier. It was through the skills, tenacity, and knowledge systems of the Nlaka'pamux people that they were still operating through management of tenures awarded to partner communities.

I was grateful to Dave for helping us design treatments informed by ancestral knowledge. We were eager to show that the management of Douglas fir forests could reflect First Nations values by prioritizing Indigenous ways of showing respect for the interconnectedness of all living things and a holistic understanding of the natural world.

3

PLANTING SEASON

Luckily, the logging company completed the harvesting just days before the snow fell in early December. When Hank emailed me the good news, I was at home in Nelson writing, staring out my upstairs office window at the old maple tree blanketed in snow between paragraphs. The tree was about the same age as my century-old wartime house, wizened and worn but built for endurance. I immediately called Jean, nestled in her home office in Kamloops, and we rejoiced on Zoom.

I could feel the building excitement that always came with a new research project. We were so close to the moment we'd been waiting for. After four years of preparation—the grant writing, planning, strategy meetings, relationship building with our nine partners, harvest prescriptions, ecosystem mapping, preharvest assessments, harvest coordination, seedling orders, seedling production, seedling lifting, storage, and delivery—we could now proceed with the most rewarding part of our experiment: going into the forest to plant trees. I had planted many experiments over my four decades as a forester turned forest scientist. I felt closest to the land when I was pouring my sweat into the trees, digging my hands into the soil, communing with the plants. Over the years, I had followed my experimental trees, measuring their size and health, whether they'd died and why, and how abundant their neighbors were. Some of my earliest plantings were now four decades old.

By spring, I was itching to get to the field to check how well the fallers had done. I started my pickup truck and headed to the bush for the first time that year. As I rounded the last corner to the spur road leading into our experimental site, the sunlight was swallowed by a looming shadow. I looked up, dwarfed by a massive slash pile the height of a two-story building. This towering mountain of remains was the logging debris left behind after the trucks had hauled away the best logs in December. Only about a quarter of a logged forest ends up in forest products. Around half is reduced to sawdust at the mill, and the rest is either scraped into burn piles or left as slash on the site. These slash piles contain more than just rejected trees: forest floor, branches with nutrient-rich leaves, and plants ripped up by the skidders are all scraped into the mix.

I parked my truck at the lowest-elevation control plot and made my way into the uncut glistening forest. A whisky jack greeted me, his gray vest fluffed, dark-hooded head cocked in anticipation of food.

Slash pile in Sinixt, Ktunaxa, and Syilx territories, 2018. Branches, tops, logs with defects, and rotten wood are pushed into piles and left to rot or burned in the fall. Piling slash is thought to reduce the risk of wildfire in the clearcuts, increase availability of planting spots, and improve browse for wildlife. Piles are often bigger than a two-story house, and when burned, can cause smoke pollution and greenhouse gas emissions. This type of slash could be used as biofuels, but hauling it to a processing facility is usually not feasible in the current economic market. This could change if carbon were adequately priced and companies were taxed for carbon emissions from logging practices.

Quee-oo, he whistled. I piped my best whistle back, then brushed past cedar boughs soft as feathers and coated in silky sheens of spring water. The braided plaits were exquisitely evolved to catch the sun and slow the spring throughfall. Neighboring western hemlocks huddled over saplings and seedlings like emperor penguins over chicks, their needles variable and spiraling to catch sun drops and to slowly release melting snow onto offspring below. This gradual trickle of meltwater from the tree crowns mediated the spring runoff, ensuring there was sufficient soil moisture to meet the needs of the awakening forest.

Dog lichens were nestled among a pallet of mosses that included lime-green electrified cat's-tail moss and red-mouthed leafy moss, whose undulating leaves twisted like a DNA spiral. The mosses were absorbing the spring water and would slowly release it in the middle of summer, helping to keep the forest moist and verdant when the midsummer drought came.

In the carpet of mosses were tenacious rattlesnake plantain and pink wintergreen. They are thought to have the marvelous ability, when growing in deep shade, to tap the photosynthate of neighboring trees by drawing it through ectomycorrhizal fungal networks that link the trees and plants together. They are also capable of fixing their own photosynthetic carbon when the sun shines favorably on their leaves. This strategy ensures enough food is available even when the light dips below photosynthetic thresholds in the dark forest understory.

These forest plants were showing me their special niches. Some could survive under the dim light at the forest floor, while others were free to spread out in treefall gaps. Each with its own genius at acquiring what it needed in a functioning ecosystem.

I ducked under the drooping hemlocks and squinted ahead for the clearcut edge, but it was still a couple hundred meters away. Stepping into a small gap with the snow just receding, I almost tripped over some yellow-flowered avalanche lilies that had been uprooted by a bear foraging for a spring meal. I checked over my shoulder. I wasn't alone here—the forest was also home to hungry bears looking for roots and grubs. "Woohoo!" I shouted, and moved the bear spray to the front pocket of my cruiser vest as extra precaution. I picked a handful of the bright yellow lilies for Mum and wrapped them in a velvety palmate thimbleberry leaf before tucking the bouquet in my vest. I peered around for bear tracks in the snow, and not seeing any, headed for the sun slanting in from the forest edge.

I stepped through the forest into the clearcut, my eyes squinting in the raw sunlight, and stopped in my tracks.

What have they done?

I'd been expecting to find a small, five-hectare clearcut patch, with a seed tree patch next to it, and a 60 percent aggregated retention treatment the next block over. That's what Jean and I had specified in our instructions to Hank. But far from being a couple hundred meters across, this clearcut looked like it was at least twenty hectares in size. It looked endless to me.

I slowly made my way into the clearcut. Fallen, misshapen trees lay crisscross over the mounds and in deep piles in the dips where avalanche lilies and wintergreens once grew. I lifted an uprooted sapling to uncover a black huckleberry, and when its slender green branches bounced upward, I noticed small buds breaking into leaf. The roots of the huckleberry were still moored in the soil, and I hoped they would endure. As far as I could see, the huckleberry bush was the only living thing, other than the new sprigs of bull thistle, white hawkweed, and prickly lettuce that seemed poised to survive here. The seed tree treatment we'd planned next to this patch had also been clearcut, and beyond it the 60 percent retention patch had been sheared. Instead of following our instructions to leave the largest trees in the prescribed densities and patterns, the loggers had cut every tree and run over every shrub and flowering plant. I reminded myself that forests were resilient, and that even though the treatment did not match our prescription, the clearcut would soon begin to heal.

I dug my phone out of my pocket and called Hank. He explained that he'd clustered three of the four clearcut replicates into a single large patch on the flattest terrain because it was more economically efficient. The fourth replicate, he said proudly, was exactly as we'd planned it, at the bottom of a steep slope. I pointed out that our map showed that four small replicate clearcuts were meant to be dispersed in separate patches across the experiment, so the clearcuts represented independent observations. He explained again that this conglomerate was best for the company. He'd swapped two of the replicate clearcuts on my map with the two partial retention treatments on this bench, meaning there was still the full suite of treatments logged.

"It didn't feel right to leave those big firs on such easy ground," he said.

I kicked the duff in frustration. The forest industry had been in an economic downturn, and I knew Hank was under pressure to clearcut rather than selectively harvest the biggest wood. His company was actually one of the most progressive in the province, making specialty products from the timbers they harvested while employing large numbers of people from the surrounding townships.

He offered that I could split the fifteen-hectare clearcut into five-hectare patches and still have all four replicate clearcuts to complete the full suite of retention treatments. When I countered that this approach did not meet the randomization requirement of the experimental design, nor would its microclimate match the other smaller clearcuts across the study, I could hear him shrug over the phone. What was done was done. Our conversation was over. I conceded that the company had been generous in agreeing to the extra work it took to log the little patches.

After ending the call, I called Jean.

"Don't worry," she said. "We'll make it work, one way or another."

We *had* to make it work. We'd already invested several years of preparation and over $100,000 collecting preharvest biodiversity and carbon data, meticulously counting every tree, plant, and soil layer in the twenty treatment plots.

I sat down on a stump and pulled out my map to try to piece together the rest of the experiment. The big open clearcut was surrounded by our retention treatments. Down the steep slopes, along the rolling benches, over the rocky knolls, in the swampy areas—most everywhere that wasn't flat with deep soil—the loggers had left mature trees in the densities and configurations we'd aimed for. The experiment was still usable, but we'd have to apply a special analysis to account for the biases introduced with the erroneous clearcut. Field experiments were just plain hard to do. If mistakes like this weren't made at the time of logging, there were no guarantees that fire, insect outbreaks, or windthrow wouldn't create even more variability in time. We would simply have to deal with the unexpected issues using creative statistical techniques.

This wasn't the first time the loggers had gotten it wrong. At one of the Mother Tree Project's semi-arid locations, the loggers had left large trees standing in the clearcut patches, having somehow confused the three replicate clearcuts we'd drawn on the map with the three replicate seed tree treatments. We'd been lucky to fix the problem with the help of a contract faller taking down the extra trees, but the incident taught us that detailed written instructions were not always enough. Loggers also needed guidance on the ground when it came to research.

It was getting late in the afternoon, and I didn't want Mum and Nava to worry, so I folded my map and headed toward the logging spur where I'd parked my truck. Crossing the boundaries of our experimental site, I made my way back toward the forest. As I approached the edge of the clearcut, I froze. Feeding on roots in the shadow of the trees was a mother bear and her cub. I squinted—black bear or grizzly? It can be hard to tell them apart from a distance. Grizzlies are larger and lighter in color and have a humped shoulder. Grizzlies also have much longer claws, but I didn't want to get close enough to tell.

The mother bear had spotted me at the same time I'd seen her. She stood up on her hind legs for a better look, peering at me as she tried to figure out what I was. I could see her straight snout and ears more clearly now. *Black bear.* I thanked my lucky stars. I stood up straighter and hollered, "Yoohoo!" She came down on all fours and hesitated. I waved my arms. It was best to appear big and aggressive with black bears. The bear and her cub turned and disappeared into the woods and so did I, in the other direction.

BY EARLY MAY, the soil at Kootenay Lake was warming and ready for planting. Jean arrived at my house the night before in her blue FJ Cruiser. She hopped out of her truck and grabbed me in a giant bear hug. "Good to see you, HH!"

I smiled at her use of my old college nickname, Homer Hog, because she said the way I rooted around in the soil reminded her of a groundhog.

Her blue eyes were sparking with excitement, her field clothes already on.

"I can't wait to get to the *busses*," she said, using the nickname for bushes we'd learned from an old Hungarian crew boss. I scrutinized my old friend more closely. I hadn't seen her since the fall and I worried that she looked pale. Multiple sclerosis had started wreaking havoc on Jean's immune system around the time I was fighting off cancer. That we had each gotten sick at the same time felt like no coincidence: Jean and I were reflections of each other. But our shared hardships had only made us more determined. If we could survive these catastrophic diseases, surely we could help solve the global climate crisis, we'd joked. Besides, I reminded myself, Jean was incredibly tough. She'd once hauled me and my backpack up a ten-meter cliff on Vancouver Island to escape the incoming tide.

The next morning, the Mother Tree Project crew—me, Jean, and eight students—converged on the landing at the top of the big clearcut. The cool spring air was electric with our excitement. I greeted Eva with a high five. She had recently started her doctoral studies, looking at the mycorrhizal connections between western redcedar, Pacific yew, and Douglas maple and how they potentially held the tree community together—a parallel to the camaraderie she had with her fellow students. Her trademark outfit reflected her frugal lifestyle as a student. Blue jeans tucked into hiking boots, and a worn long-sleeve shirt over a UBC Forestry T-shirt. She had plaited her hair into two braids to keep it out of the way while she planted.

The back of Jean's truck was packed to the gills with thousands of pigtails—foot-long wire stakes with a loop at the top—to which she'd tied flagging tape of different colors uniquely cross-referenced to the different types of seedlings we were planting. There were green, blue, and pink for the different species of trees—western larch (*Larix occidentalis*), ponderosa pine (*Pinus ponderosa*), and lodgepole pine (*Pinus contorta* var. *latifolia*). Most importantly, for interior Douglas fir, there were seven different patterns and colors of flagging for the seven genotypes.

Seedling genotype was the crucial third factor in our experiment. This was Jean's masterpiece, and with my postdoc Brian's help, it had taken a herculean effort to plan. The forests at our inland sites were dominated by interior Douglas fir (*Pseudotsuga menziesii* var. *glauca*), a variety with waxy blue-green needles protective against arid summers. At our coastal forest, *Pseudotsuga menziesii* var. *menziesii* predominated, its emerald needles able to rapidly photosynthesize in the lush maritime climate. At each of our nine forests—at what we can think of as their climatic locations—we were planting seven different genotypes of Douglas fir. In addition to the locally adapted genotype—the one that ought to do best if our climate remained relatively stable—we were planting three genotypes transferred from increasingly drier and hotter climates, and three from increasingly cooler and wetter climates.

Looking closely at the Douglas firs, freshly delivered from the local nursery, I could see the pale green seedlings had originated from arid locations and the blue-green seedlings from more humid climates. But with so many gradations in between, they would be impossible to tell apart once they were in the ground. As we planted, according to Jean's careful field planning, we were to insert the appropriate color-coded pigtail in the soil alongside each seedling.

The performance of these special seedlings could show us how to assist migration of seed sources in order to supplement natural regeneration in our forest plots, creating rich mixtures that could adapt to the rapidly changing climate. Seedlings in our neck of the woods usually performed best when planted close to the environments of their seed mothers and pollinating fathers. This was because the strongest natural selection pressures that cause adaptation in genes occur where the seedlings originated. Not unlike humans, trees are deeply influenced by their families and local neighborhoods. Temperature fluctuations, rainfall patterns, local soil condition, the suite of predators, and the neighboring trees and plants, to name a few, all exert selection pressures on the local genotypes, as expressed in the trees, in a forest. But based on years of accumulated knowledge, foresters also know that some tree seedling populations are well adapted to a broader range of conditions than just where they germinated.

Moving seedlings away from their place of origin, or natural range of dispersal, can be quite beneficial in cases where danger has moved in. Changing climatic conditions or pathogen loads around parents, for example, can kill seedlings at the time of establishment, whereas moving them farther away could help the seedlings survive. Seed transfer, also known as assisted migration, is a tried-and-true technique that has been around for centuries. Douglas fir has been widely transplanted around the world, from its natural range in western North America to places as far away as Scotland and New Zealand. Within its natural range from Mexico to central British Columbia, people have also begun cautiously moving genotypes northward to help keep pace with climate change. By comparing the performance of seven genotypes at each of our climate locations, we could determine from where and how far Douglas fir genotypes could be migrated.

Jean and I and our group of students would start planting at our southern sites near 49° latitude, then gradually work our way northward following the pattern of snowmelt. Our plan was to finish six weeks later, in mid-June, at our coldest, northernmost location, John Prince Research Forest north of Fort St. James, just below 55° latitude. If our timing worked just right, the last patches of snow would be disappearing as we dug the first trees into the ground at John Prince.

Jean held our tailgate safety meeting, then explained the trees and the pigtails. Beaming, she pulled waterproof tables and maps from her exquisitely organized three-inch binder and handed copies to each of us. She organized us into two crews of five, each group to tackle one measurement plot at a time. With one larch, two pines, and seven genotypes of Douglas fir in our suite of seedlings, each planter had two types of trees to look after per plot. Jean instructed each of us how far apart our particular species or genotypes needed to be planted—the distance for each unique type according to her master plan—then showed us how to zigzag the plot and move around each other so we would create an intimate mixture. We were to plant twenty-five thousand trees here at Kootenay Lake, then another two hundred thousand trees at the rest of the locations. Once our experimental trees were in the measurement plots, a commercial planting crew would follow us to plant the surrounding buffer areas using trees prescribed by the logging company as required by their harvesting license. With the company trees growing in the buffer zones, the entire forest would be stocked at the same time, providing good neighbors to our test trees. The boxes of trees stacked, we loaded our planting bags with bundles of seedlings that looked like sturdy carrots just plucked from the garden.

Tall, lanky Liam, his blond hair tied back in a ponytail, was the first to put on his planting bags, suspenders battened down, so he could load the three seedling compartments. Liam was our longest-running crew member, having started when he was seventeen years old. Over the years he had become an expert technician and crew leader, teaching the other students the methodologies. Liam jumped into the back of Jean's truck and started passing the boxes of trees, two at a time, down to the others. The rest of the crew quickly formed a fireman's brigade to stack the boxes in the shade of the forest near the creek.

I packed eight bundles of western larch and ponderosa pine into my bags, along with their green and pink pigtails. I loved the weight of the bags, my shoulder straps cinched up to keep some of the load on my hips, planting shovel in hand. My shovel was about the length of my arm, just under a meter of solid metal, perfectly weighted and shaped for swiftly slicing the earth. The blade ended in a rounded point that could pierce through duff, twigs, and roots with the least amount of effort. I balanced it in my hands like a baton. I was eager to put the trees in the ground, the finishing touch to launching our experiment.

My first shovel went into the ground like a hot knife in butter. The next spot was half a meter deep in slash, so I swept the branches and twigs away and plunged my shovel blade into the earth. *Clank!* I'd hit a rock, the vibrations shooting up my forearm. I tried another spot half a meter away. *Twang!* Another rock. I swept away more slash, looked carefully at the contours of the earth, then found a sweet spot between the stones. I needed to read the soil better, protect my tendons, conserve energy, and maintain speed. I walked forward a few paces, watching where the fireweed was poking up as a sign for good soil, then plunged my shovel plumb down next to a stump that could shade my seedling. *Niiiice!* The silty loam was perfect for the larch I tucked into its home.

I imagined this beautiful soil thrumming with life. The soil food web working like a relay race, the big centipedes and arthropods shredding the twigs and leaves, the smaller millipedes and collembola chewing the fibers and petioles, and the smallest springtails and mites slurping down the bud scales and leaf veins. Each critter was passing the baton—a smaller bit rendered from their consumption—for the next one down the food chain. These shredders and chewers would reduce the detritus to a size where the amoebae, nematodes, and gastropods could get to work on the clusters of dead cells. Each of these creatures would also feed on soil organisms that were smaller than they were, releasing what they couldn't absorb, and in so doing cycling the litter until it was a colloidal concoction of excrement and cytoplasm. Then the microorganisms—the fungi and bacteria—would reduce the remains into an even smoother soup of nutritious compounds that the plant roots and mycorrhizal fungi could readily absorb.

Our seedlings would use the goodies in this soil solution to synthesize the ATP and NADPH molecules needed in photosynthesis and respiration. These energy-rich molecules are used to fuel the creation of glucose and other carbohydrates for the next stage of photosynthesis, called the Calvin cycle. They'd make the proteins, enzymes, and nucleotides essential to tissue construction and DNA synthesis, and the carbohydrates, cellulose, and lignin needed to build leaves, stems, and roots.

I imagined the whirring and chirping of photosynthesis—the light reactions of photosystem I, the electron transport chain, and the dark reactions of photosystem II—to transform the sun's energy into the chemical energy needed to synthesize the carbon molecules downstream in the Calvin cycle. I could almost hear the melodic notes and bars of respiration as the sugars were metabolized—using the processes of glycolysis, the Krebs cycle, and phosphorylation, together playing like a musical score—so the plants could use the energy to carry on the business of life, taking up more water and nutrients, making new tissues, synthesizing defense molecules, and packing on biomass. The soil food web and the seedlings would turn the dead slash and fallen litter into new life again. As the sun rose higher in the sky, I imagined the rate at which these creatures bit and chewed and ate increasing as the temperature of the soil nudged upward.

By the end of the first day, I was dead tired. Looking over the measurement plot our little crew had managed to plant, I was happy to see Jean's colorful pigtails sprinkled randomly about like confetti.

After Kootenay Lake, we planted the Jaffray site near Cranbrook, deep in the Rocky Mountain Trench, its climate semi-arid and the Douglas fir forest less productive. Only 300 kilometers (185 miles) from Kootenay Lake, we'd had to cross the Selkirk and Purcell mountain ranges to get there, and with so much rainfall in the cooler mountains, the air had dried out by the time it hit the Trench. We drove there in our caravan of student rattletrap cars, crew-cab pickup trucks, and my dad's 1990 Chevy Silverado, its exhaust system as loud as a Harley-Davidson.

It was hot and dry by the time we arrived at the Jaffray forest, the wood ticks jumping from junipers and rattlers dozing in rockpiles. I worried how our seedlings would fare among the pinegrass under the trees and in the thick domestic grass clumps in the clearcuts, carried there by cows and ranchers. It was well-known that grasses like these competed with seedlings for soil moisture. By the middle of the first day, we were all sweating heavily, downing liters of water and resting in the shadows of our pickups. Unlike at forgiving Kootenay Lake, I figured our seedlings would be under extreme climatic stress and would survive better under the shade of the mother trees than in the sunbaked openings.

The cool weather had returned by the time we went back to the West Kootenays to plant Narrows and Two-Bit, but this wasn't entirely unexpected given that these sites were nestled in the rainy mountains alongside the mammoth Kootenay and Arrow Lakes. When we arrived at Two-Bit at the end of May, the fireweed was getting tall, the deep loamy soil was moist from the rains, and the woodpeckers were hammering away at dead trees. The year before, we had lost two of our replicates at Two-Bit to a windstorm, leaving behind some snags for the cavity-nesting birds, but not enough live trees to carry out our treatments. Luckily, a single replicate had been left unscathed.

Six hundred kilometers (375 miles) north of Nelson, we made our way to the University of British Columbia Alex Fraser Research Forest, where one of the replicates had burned the year before. Now it was brilliant with magenta fireweed, the dense plants covering the charred soil and returning life to the forest. We plugged our seedlings into the logged openings big and small, and under the trees scorched and alive, finding the soil warm and moist where our seedlings would surely thrive. Our replicate site at the lowest elevation at Alex Fraser was less forgiving, with heavy clay soil and swards of grasses, and I worried that the trees would be entombed in solid ceramic when summer came into full swing.

The next day, at the wettest of the three Alex Fraser replicate forests, the rain fell so hard our planting bags flooded with water, and my glasses fogged so badly I had to remove them and plant blind. But we knew our seedlings were loving it, so we were happy.

WE ARRIVED AT JOHN PRINCE, our final, northernmost site, in mid-June, just as the snow was disappearing from the woods. The John Prince Research Forest, just northwest of Fort St. James, was situated on a rolling plateau of spruce, pine, Douglas fir, and aspen within the territories of the Tl'azt'en, Binche Whut'en, and Nak'azdli Whut'en First Nations. The nations collaboratively managed the huge, 16,000-hectare (40,000-acre) landscape with the University of Northern British Columbia.

By the time we got there, we'd weathered weeks of long hard days, blistering sun, torrential downpours, tick bites, and endless pizza and burrito dinners. The end of planting was tantalizingly close, so we raced through the plots, trying to keep up with Kaya, who was one of the fastest planters on our crew. While planting in one of the 60 percent patches, Kaya stumbled across a bear den, a grassy bed nestled under some logs in a cluster of aspens, with deer bones littered around the entrance. The den appeared empty, but large grizzly scats suggested the bears were in the area. We gathered as a group and quickly planted around the den, calling "*Woohoo*" to warn the bears off until we were finished. A bear biologist later confirmed that clearcutting would have made the den unusable, leaving the mother bear without this place to hibernate and give birth that winter.

We were down to the last few plots in the upper block and the rain was turning the road to mud. A lake had formed at the base of a crucial spur leading us to the plots. Our energy drained, no one wanted to carry the boxes of trees to the upper slopes, so instead we carefully navigated our vehicles through the mudhole. We planted the last trees as the sun was setting, and we cheered and danced and pulled down the tailgates to celebrate with cans of lukewarm beer.

"Are we done, sir?" Eva asked. I had discovered that a rich vein of playful humor lurked beneath Eva's serious exterior. For weeks we'd been goofing around, driving the rest of the crew crazy as we assumed the characters of Sherlock Holmes and Watson and exclaimed loudly in English accents how much we "*loved* nature."

"Yes. Good work, Watson," I answered. Everyone groaned, then doused us with the dregs of their beer.

"Time to go, Watson!" Eva and I jumped in the Silverado and slammed the doors shut before the others could catch us and cover our mouths with duct tape. We rolled down the road back toward the camp, my eye weighing the mighty mud puddle, realizing it had grown larger since morning.

"Gun it, sir!" Eva said, and I put my pedal to the metal. We slipped and squealed and hit the puddle full speed, only to grind to a halt in the middle of the swamp.

"Watson!" I cried, and Eva jumped out up to her knees in the mud.

"Oh dear, sir. I do believe we're stuck."

"Stand back, Watson!" I threw the truck into reverse and gunned it again. The wheels spun deeper.

"Stop sir, this isn't helping." Eva's clothes were spattered with mud. I opened my door to find mud up to the floorboards.

The other students gathered in the muck behind the truck and started pushing. I hit the gas again, but we only ground down deeper.

Everyone started digging out the wheels with their planting shovels, while Kaya laid some saplings behind the tires for traction. More digging, more saplings, more pushing. An hour later, we were still stuck.

"We need a tow rope," Jean said. Of course, we didn't have one, though we had everything else to get out of a pickle like this.

We managed to drive the other trucks through the mudhole and back to the camp just as it was getting dark. There was a tow rope in the shed, and after beer and nachos, anything seemed possible by the time we went to bed.

The next day, the wheels of Dad's poor truck had sunk even deeper

Members of the Mother Tree Project crew, left to right: Hannah Sachs, Alexia Constantinou, Curtis Rock, Ari Murphy-Steed, and Eva Snyder. The crew were enjoying lunch after completing a National Forest Inventory plot, measuring biodiversity and carbon stocks, in Nlaka'pamux te Secwépemc territory near Peter Hope Lake, British Columbia, prior to the experimental harvesting. This preharvest data would provide a baseline for tracking the impacts of harvesting on the ecosystem over time. Experimental approaches with both preharvest data as well as uncut controls for comparison are known as Before-After-Control-Impact (BACI) designs.

into the mud overnight, and we were worried the other trucks wouldn't have enough torque to pull the Silverado from the pond.

"Pull the tow ropes tight first," Jean said, after we hitched the Silverado to the blue pickup. I shoved the Silverado into neutral and Kaya inched the crew cab forward until the tow rope was taut.

"You can do it, sir!" Eva grinned at me.

"Go!" Jean brought her arm down like the starter at a car race. Kaya hit the gas of the blue truck, her wheels spinning on the gravel road. I felt the Silverado creak and groan, then inch forward. I shoved the transmission into low and the Silverado spun out of the mudhole onto solid ground, everyone cheering.

I was so proud of our crew and what we had accomplished thus far, and so gratified by the trust that had developed among us. Trust is the essential ingredient for any collaboration. While planting our trees, we had shared our hopes and nurtured our relationships, the back-and-forth and give-and-take, and this reciprocity had bound us together into a thriving human ecosystem. Each individual on our crew, whether new or experienced, brought strengths and weaknesses, but we all worked together, helping one another and contributing something unique and valuable to the community. Likewise, Jean and I worked in symbiosis on our shared projects, me keeping my eye on the big picture and Jean filling in the details.

We left John Prince tired and happy, hoping our trees would take to the soil over the summer. As we piled into the trucks, Eva shouted that an ember had just landed on her phone and melted a hole in the screen. Glancing at the sky, I saw a solid orange band across the horizon to the north. I turned on the radio to listen to the news. The province was under a heat warning and wildfires were igniting to the south and north of us. As we turned onto the logging road heading out of the camp, I spotted a mama grizzly and her two cubs watching us drive away, unaware that a climate disaster was brewing.

4

KILLER FORESTS

In September, Jean and I returned to the scorched forest at the Mother Tree Project site in the Cariboo Mountains, where we'd been chased out by wildfire months earlier. The smoke of 1,353 wildfires burning in the interior of British Columbia had spread across the province, even drifting over the Coast Mountains to the sea.

On our way north, we drove through the Elephant Hill fire in Secwépemc territory. This fire would eventually burn through more than 1,900 square kilometers (730 square miles) of forest, woodland, and grassland, an area about the size of Liechtenstein. The fire had leaped a river, raced up mountains, and skipped over firebreaks, and was still not out when we puttered down the hill into the tiny town of Cache Creek on the fire's southernmost flank. Thanks to the hard work of firefighters, the town had been spared the flames. I pulled up to the only four-way stop, the red light flashing eerily through the thick smoke. If anyone was on the street, we wouldn't have been able to see them.

I knew this town well. When I was fifteen, in 1975, I'd worked two part-time jobs here: as an apprentice lifeguard at the local swimming pool, and as a weekend cowhand for the Bonaparte Ranch, where I'd changed sprinkler pipes in the hay and alfalfa fields bordering the Bonaparte River. The father of the Secwépemc family I'd boarded with had once taken me flying in his little two-seater propeller plane over the grassy silt cliffs along the meandering Bonaparte and its main stem, the swirling Thompson River. The oxbow Bonaparte River had flooded in 1894, and again in 1948, but otherwise its banks were ever adjusting to absorb the spring freshet. He'd let me pilot the plane for a moment, soaring over the fingers of ponderosa pine and Douglas fir that rose up through creek chasms, and above the lofty spruce and subalpine fir forests that spread over the highlands.

There had been little logging in the surrounding forests back in 1975. But over forty years later, Jean and I could see the landscape had become an ocean of clearcuts. Much of the primary forest has been logged, sometimes with salvage logging after beetle outbreaks or other fires, but mostly for green timber. Now the slopes surrounding the town were charred black.

"I bet Cache Creek will flood." I pictured gravelly water rushing over the paved four-way intersection.

The floods had become more frequent in recent years. There was one in 1979, blamed on the development of a subdivision, and another flood eleven years later due to highway and bridge construction. After a twenty-five-year hiatus, the floods returned in 2015 and 2017, this time attributed to torrential rain and rapid spring runoff. At the time of the 2015 flood, an alderman noted that the drainage system had been designed for normal precipitation patterns, not flash floods. He said, "There's no way you can design a system that would take a thing like that; the water has to go somewhere."

We would later learn that the Elephant Hill fire had seared more than half of the Bonaparte River watershed, burning so intensely through the mature forests and clearcuts that in places it killed not only the trees and plants but also the duff. Many of the seeds in the soil were destroyed and the lodgepole pine cones that would normally open with heat were baked, resulting in spotty natural regeneration of trees or other plants after the fire.

Where the forest floor wasn't turned to ash, it was scorched into a waxy, hydrophobic crust. This water-repellent layer prevented the soil from absorbing the rain when it finally came in the fall of 2017, causing erosion and sliding into the creek beds. That winter, the snow accumulated to greater depths than usual because there were no tree crowns in the burned and clearcut forests to intercept the blizzards. When the hot sun hit the heavy snowpack in the spring of 2018, the charred soils couldn't absorb the rapid snowmelt. A torrent of water

Forest burned in Elephant Hill wildfire in St'uxwtéws te Secwépemc territory, 2017. The fire burned so deeply in places that walking through the ash was like wading through knee-deep snow. The dark furrow on the right is the remains of a creek.

triggered landslides that filled the Bonaparte River with sediment, rocks, and logging debris, eroding the banks and flooding the alfalfa and hay fields where I'd once worked. The elementary school near the swimming pool was evacuated the morning of April 27, 2018. By early May, the whole town had flooded.

The village of Cache Creek commissioned a flood mitigation plan in 2021. The engineering firm noted that the Bonaparte watershed already had a "relatively high equivalent clearcut area" prior to the Elephant Hill fire, and when the rest of the high-elevation forests subsequently burned, this cleared area rose to "very high," and with it a "high severity flood hazard." The government at the time noted that high flood risk after wildfire can last up to five years, until the forest starts to grow back, but risks may amplify with climate change because grasslands are now replacing forests while wildfires are expected to become more frequent. They recommended replanting clearcuts and burned areas "using a science-based approach to reduce flooding."

No mention was made of curtailing further clearcutting. The Bonaparte has flooded almost every spring since the Elephant Hill fire.

AS WE WOUND OUR WAY northward through the ashes, Jean's face gradually folded in dismay. Spectral wisps of smoke clung to blackened sagebrush and ponderosa pine, two species she particularly loved. The air smelled acrid and left a gritty film on our skin. The landscape felt like a graveyard. No birds sang, no leaves trembled in the breeze. Ground squirrels, badgers, and burrowing owls had surely met painful deaths. Creeks were choked with soot. Glancing over at a smoldering shrub, I felt a heaviness in my chest, knowing a million wild voices had been silenced.

We knew fire was healthy for ecosystems and helped maintain balance. It reduced fuel loads, released nutrients, enhanced regeneration of early successional plants, and created habitat for certain species. Aspen, strawberries, fireweed, scarlet gilia, pinegrass, liverworts, and morels. Moose, mule deer, and coyotes. And the pattern of fire behavior was usually a mixed bag: flames would skip over hollows of moist forest and peter out along rivers, scorching but not killing thick-barked Douglas firs and ponderosa pines, melting the resins of serotinous cones of lodgepoles to release pine seeds, and stimulating aspens to shoot up suckers from their clonal root systems. But a sense of dread hung over us. The three ingredients essential for fire—heat, fuel, and oxygen—were all being dramatically affected by climate change. The amplification of the trifecta was causing more severe and extensive fire across our forests than ever before.

In early July 2017, multitudes of lightning strikes had ignited massive dark smoke columns, launching a season of superheated, out-of-control wildfires. A dome of extreme heat had set up early in the spring and persisted in the absence of rain for weeks, sucking the moisture out of the forest. These heat domes had become more common in recent years, and the heat this summer had also generated more lightning strikes than usual. Hotter temperatures increase lightning activity because they result in more water vapor, which creates an electrical current when it rises, condenses, and the ice crystals collide in thunderclouds.[1]

When the first bolt of lightning struck on July 7, the forests were already vulnerable, not just due to the drought, but also because their understories were choked with tinder-dry fuels: desiccated brown twigs and needles, grasses, and leaves. This was partly because of the government policy to suppress fire, not only to protect the ever-expanding encroachment of human settlement into forests, but to save the forests for commercial timber or other interests.

About 60 percent of fires in British Columbia that summer were ignited by lightning, the rest attributed to good old human error—a spark from a vehicle, an unattended campfire, arson. The Elephant Hill fire had been started, it would eventually be discovered, by "smoking materials"—a discarded cigarette or match tossed on the ground.

As I peered through the windshield at the smoldering clusters of trees, I could easily piece together how the flames had climbed from the understory layers of suppressed trees into the crowns of the elder Douglas firs. Once the fire was high in the tree crowns, there was no stopping it from racing across the landscape.

Not only had the mature forests burned, so had the young managed stands.

"These plantations are toast," I said as Jean and I passed one incinerated clearcut after another. In wide swaths where mature trees had been logged and replanted with even-aged monocultures of lodgepole pine saplings, black charred sticks were all that remained. In the lower stretches alongside creeks, scorched stems and blackened duff indicated the fire hadn't burned quite as intensely, but these parts of the plantations were still severely damaged.

As the forest industry increasingly liquidated primary forests to provide wood products and fuel the economy of British Columbia, extensive areas of native forest had been converted to single-species plantations. Although the trees planted were native to British Columbia, these tree farms were not much different than the commercial eucalyptus plantations—replacements for the native oaks, laurels, and chestnuts—that had ignited and burned in the hills of Portugal in June 2017, killing sixty-six people and injuring more than two hundred others. Eucalyptus had been planted extensively in Portugal's countryside to feed the powerful paper industry, but the trees' oils, sap, and peeling bark burned fast and hot. The burning leaves and bark of the eucalyptus trees were carried by the wind, spawning dozens of deadly infernos. Vital Moreira, a Portuguese law professor and former member of European Parliament, had become an outspoken critic of the "uncontrolled invasion" of eucalyptus that now covered a quarter of his country's forest land. He had written, "It's not enough to cry for the dead.... We have once and for all to deal with this powder keg represented by the forest we chose to have. We have to recognize that we've created a killer forest."[2]

The conifer species planted here in British Columbia were similarly chosen because of their rapid early growth rates, so they could establish quickly and provide commercial timber again within the shortest possible time frame, but they were also highly flammable due to the large amount of sap in their branches. To boot, the needles of these species had high resin loads and low water content, so they burned easily.

I could see that the charred saplings were crowded together, planted at close spacing to increase the volume of wood produced, which would have promoted the rapid spread of the flames. One burning branch torching the next. The blazing limbs would launch like projectiles, seeding new flames across the plantation, and to the next one down the road.

I also knew that any trembling aspens or paper birches would have been removed from these commercial tree plantations with herbicides or brush saws because they were considered competitors to the conifers, a common practice meant to ensure that the stand was fully stocked with lucrative trees for the next harvest. With their high foliar water content, the deciduous trees acted like natural firebreaks in the forest, but this ecological buffering against wildfire was of lesser interest than the financial gains expected from manicuring these plantations. Our forest management practices had created a province-wide stack of kindling, just waiting for a spark to come along.

"You can see where the fire slowed at the aspen groves that are outside of the logged area." Jean pointed up the hill, and I felt a familiar sense of relief at the sight of the gleaming white trunks and quaking green leaves. The paper birches Granny Winnie grew in her yard in Nakusp had fortified the gardens and helped protect their wooden house from the threat of fire. After a wildfire scorched the hills above Nakusp in 1973, the early successional birches dispersed prolific, lightweight seed and sprouted quickly, speeding up recovery of the surrounding forest. The deciduous species also protected the conifers regenerating in the blackened soil from the extreme soil surface temperatures. The rapidly recovering forest following the fire lessened the risk of flooding when the Kuskanax River flowing through town rose the following spring.

I had spent the past three decades trying to convince the provincial government that aspens and birches were essential to forest vitality—efforts met with stonewalling. An all-out war on deciduous trees had gradually diminished these crucial firebreaks. The mountains and plateaus had become a sea of coniferous plantations, akin to a carpet of matchsticks rolled out for the flames to race across. These plantations were still broken up with primary forests, but those, too, were tinder-dry with surface, ladder, and canopy fuels after fire suppression policies had interrupted epochs of Indigenous fire stewardship. Adding to this mix were the millions of hectares of trees that had been killed or weakened in the previous two decades by outbreaks of mountain pine beetle, Douglas-fir bark beetle, and spruce bark beetle. All of these infestations had been amplified by climate change. Extensive salvage logging of the dead trees had emitted an additional pulse of greenhouse gases and increased the homogeneity of the refurbished landscape. This had left the terrain even more vulnerable to catastrophic wildfire.

BY THE CLOSE of the year, 2017 would clock in as one of the most destructive fire years on record in British Columbia, with over 12,161 square kilometers (4,700 square miles) burned. Scientists calculated that eleven times greater area burned in British Columbia than expected. This record would be beaten again the following year, and again in 2021, and again in 2023, also smashing records for temperature. In 2023, the most destructive wildfire season in Canada's history as of this writing, megafires would burn an area three times the size of Nova Scotia across Canada.[3] Thick smoke would choke the continent, drifting as far as New York City, turning the skies orange and triggering hazardous air quality alerts.

In recent years, the fires had started burning earlier in the spring, even before budbreak and leaf-out, and the snow was melting more rapidly from the extensively clearcut and burned landscape. In the absence of green, moisture-laden leaves to dampen the flames, these early season fires were especially destructive because they burned deeply into the forest floor, destroying roots, seeds, and soil organic matter and leaving, in some places, ash that was knee-deep. On steep slopes, when the rains came, the ash started to run, gathering up soil and rock, and often leading to landslides. Wildfires had raged through the summer and fall, sometimes continuing to burn right through the winter. Under the snow, the fire smoldered in the duff and roots and erupted again in the spring when the soils warmed and the winds picked up. These overwintering fires, known as "zombie fires," would smolder even through bone-chilling blizzards.

When Jean and I reached the logging road leading to our experimental plot, we could see red lights flashing. An officer jumped out of his truck, waving his arms and telling us we couldn't pass through because the fire was still active. I could see smoke rising above the canopy and my heart sank knowing the trees were even now burning. I looked over at Jean, her face aglow from the red strobe light.

"It will be all right, HH," she said, in her signature Lucy-the-psychiatrist voice. It was a long-running joke between us: Jean pretending she was Lucy, giving psychiatric advice to me, Charlie Brown.

"Some of the trees will survive."

AT HOME IN NELSON the following spring, my qualms about wildfire were overtaken by my burgeoning anxiety about Mum, who had been diagnosed with dementia a couple years earlier. Mum had always been the backbone of our family, listening to our triumphs and tragedies, giving us sage advice, and encouraging us to do our best. She'd taught us about women's rights, social justice, and the need for a good education. She'd shared with us her delight in wild places, her beloved lakes, shores, and mountains.

Now Mum was losing her mind. Alzheimer's was the one thing she'd feared most after losing her own mother, Granny Winnie, to the same disease twenty years earlier.

I had just finished washing the breakfast dishes when I heard a knock at my back door. I looked through the glass pane to see Mum, hunched over, walking stick in hand—a different one than the pole she'd left behind yesterday. At eighty-two, Mum's white hair was short as chickadee feathers, her body fit from hours spent each day tending her garden. But today her pale blue eyes were filled with anxiety.

"Suzie, the fires—should we go to the lake now?"

I helped Mum over the door sill as I glanced past her shoulder at the sun, seeing no sign of the smoke that had darkened the sky the summer before. Her black cat, Kitty, was sitting nearby, having followed Mum from her house three doors down the alley as he always did. Kitty fixed me with his green eyes, making sure I was there to help. Kitty adored Mum's gentle laughter and kooky personality as much as I did. I savored memories of sipping Coronas on the front porch while dinner simmered on the stovetop and Willie Nelson played on the cassette deck. Mum always made everyone feel at home.

"There are no fires yet," I reassured her, as I sat her down at my kitchen table. "It's too early in the spring." The stove was warming the house and the rich aroma of coffee filled the air. The two cups I had washed were ready for me to fill. I wanted more than anything for Mum to be well, to live forever, and it hurt to see her so anxious. I didn't let on that the fires had already flared up in the north due to the unusually warm, dry spring, igniting even before the aspens and grasses had leafed out.

Nava helped Mum shed her yellow raincoat and knee-high red gum boots, but didn't mention that Mum had her shirt on backward. Nava was at home finishing grade twelve at LVR Secondary School, while Hannah was in Vancouver completing her second year in the Faculty of Forestry at the University of British Columbia. Hannah had started at university for a degree in kinesiology, but after a year realized that her heart was in the forests.

"I get why you're nervous, Granny." The close call we'd had in the forest had left a vivid impression on Nava. She'd seen firsthand how terrifying wildfire could be.

Since my daughters had been born, the area burned by wildfire in British Columbia had been rising exponentially.[4] I recalled my distress in 2003 when flames had scorched the forests I'd worked in for decades and when homes in Barriere and Okanagan Landing were torched. On our visit to Mum in Nelson that summer, we'd planned our route to avoid the smoke choking the interior towns.

But that year of fires paled in comparison to 2017, when 1.2 million hectares (3 million acres) burned. My daughters' summer bush jobs had been shut down, forcing them inside. In 2018, the fire season was again off to a quick start, with blazes already burning across Canada. Mum was tuned in to the national news on CBC, and she was right to worry.[5] This fire season would eventually smash all of last year's records.

Nava opened the door to let Kitty inside and he immediately jumped onto Mum's lap, purring and kneading. We'd all been stressed by the fires over the past few years, but Mum had been deeply affected. While her cognitive decline had been slow, threats like fire created an overwhelming fear in her. The 2015 Sitkum Creek fire up the west arm of Kootenay Lake had come within four kilometers of her house, and we'd all watched as the Martin Mars water bomber scooped water from Lakeside Park to prevent the flames from spreading into town. She had nightmares of wildfire racing down the mountains and kept a packed bag in her kitchen for a quick escape. She fretted that Kitty would get left behind.

Mum started to relax as we bantered about what to include in our escape kits—food, water, prescriptions, wallets, insurance papers, keys, phones, camping gear, flashlights. A full tank of gas.

"Don't forget food for Kitty," Mum piped up, her survivor's instinct as sharp as ever. I showed her the clear sky outside and her anxiety dissipated.

Nava soon set off on foot for school, and Mum and I took a tour of my garden. The snowdrops were in full bloom, their delicate white, bell-shaped blossoms framed against sturdy dark green basal leaves.

"Spring is here, Suzie," Mum said, holding my arm while using the tip of her walking stick to scrape away the remaining patch of snow covering the littlest blooms. Granny Winnie had planted snowdrop bulbs under the paper birches in Nakusp, and I remembered how, as kids, we'd find Easter eggs hidden around the drooping clusters of flowers in the spring.

"And here is Nava's shrub," Mum said, using her stick to point at the forsythia, its early blossoms brilliant yellow against the pale spring sky. She remarked on this plant whenever she came over. I remembered Nava planting it the day we moved into the neighborhood just after my husband, Don, and I separated in 2012. I'd chosen this house so we would be close to Mum in her later years. Don and I would divorce the following year, but he would live only minutes away, allowing us all to spend time together regularly. Even though Don and I were apart and he was seeing someone new, we still spent holidays, birthdays, and many weekends together. The girls stayed with him whenever I was away, and lived with me when I was home.

After visiting the dormant gingko and awakening grapevines, the green-budded lilacs and catkin-festooned hazelnuts, Mum and I made our way down the alley to her house, Kitty following a few paces behind. My older sister Robyn and her daughter, Kelly Rose, arrived just as we were settling into Mum's rocking chairs with the remains of our coffee. Robyn, her long dark hair clipped back in a bun and yellow cotton dress swishing against her leather boots, arranged a bouquet of Easter lilies in a vase. She was on her way to teach grade one at Rosemont Elementary School. Robyn had taught generations of kids in Nelson, and her teaching wisdom was well-known around the school district. Kelly Rose sat cross-legged next to Mum and pulled out her knitting. She was in the middle of her psychology degree at the University of Victoria and was home for Easter break. She asked Mum how Kitty was doing, knowing it would prompt stories of hanging out with the alleyway skunks and scuffling with the raccoons.

"The chicks have hatched," Robyn said, placing a crocheted doily under the vase as she described how her students had set eggs under a heat lamp and watched as the tiny beaks poked their way out of the shells. Mum laughed, having shared these classroom lessons with Robyn through the years.

"Did any of the chicks escape?" she asked, her eyes mischievous, imagining the kids running around catching the yellow feather puffs.

"Oh shoot!" Robyn looked at her watch. She gave Mum her pills from the blister pack and then slipped out the back door to head to her classroom. By then, Mum had forgotten about the fire escape bag and was getting ready to weed the garden with Kelly Rose. I followed Robyn to her bike so we could go over the events of the past couple of weeks.

"Mum had an anxious morning," I said. She'd lost her purse, found it in the fridge, then set off the fire alarm in the middle of the night, summoning the fire department.

To our surprise, Mum had actually improved cognitively since receiving her diagnosis of Alzheimer's. But on the front lines, the signs were clear that she was struggling. Robyn and I agreed that together we would look after Mum so she could stay in her own home as long as possible.

A YEAR AFTER we'd run from the flames, we were finally able to return to our Mother Tree Project site near Likely. I was filled with uncertainty about what we would find there. Hannah and Nava led the way through the seared forest as I followed slowly with Jean, every so often poking my shovel blade into the duff to gauge how deeply it had burned. A breeze picked up as we walked. The lime-green liverwort was already spreading over the blackened earth. Rosettes of flattened thalli and umbrellas of star-shaped gametophores stood brilliant against the darkened mineral particles and charred duff. The diminutive nonvascular plant had begun its healing work, turning sunlight into sugars, soil particles into aggregates, and litter into humus. My heart lightened. The liverwort was preparing the soil to absorb the rainwater and provide a foothold for lichens and mosses to establish.

In the charred remains of the duff, I could see that new Douglas firs had germinated between the liverworts while lodgepole pine cotyledons emerged next to opened cones. The radicals of these seedlings plied the riches of opened soil pores and absorbed the nitrogenous soup that had been released during the combustion of proteins and nucleic acids. In the sudden death from fire, life here had continued in constant motion. Bursts of fireweed, with its racemes of magenta flowers and bright lanceolate leaves, flowered among the charred Douglas fir trees. The plants were grass green, replete with the nitrogen, sulfur, phosphorus, and other essential nutrients that had been liberated from carbon lattices with the combustion of the forest floor.

The elder trees, it seemed, had for the most part survived the fire. In the understory of these hardy survivors, saplings and shrubs were burned to the bone, but a lone willow splashed pale green against crisp needles. A sprig of buffaloberry and a sprout of sticky laurel burst viridescent around a burned stump. New life was building around old life. The ecosystem was reorganizing around the legacies left behind. We climbed the knoll to where we'd started running from the flames a year earlier, holding our breath at what we might see.

"She's alive!" Nava broke into a run, then threw her arms around our beautiful mother tree, blackening her cheek as she pressed her face against the tree's rough bark. Candled flames had seared the tree's skin and burned her lower branches. Her thick cork was blackened and the deep vertical fissures carbonized. Nava planted a kiss on the wizened giant. She turned to me with a smile tinted in soot and laughed with such delight that an owl hooted nearby. We all looked up. High above, the tree's crown was bright green with flushing shoots. Soft, feathery needles coating new twigs. Her large branches had remained muscled and were once again alive with bright pink female cones. Red squirrels chattered in anticipation of the bounty of seeds to come. A black-capped chickadee peeked from an abandoned woodpecker hole, and sang, "*Chick-a-dee-dee-dee.*" I was filled with gratitude at being here in this forest with my daughters, thankful for the lessons it was teaching us about resilience and regeneration.

Hannah was eager to show me that most of the canopy trees in our plot had survived. She led me to the thickest grove, where the neighboring trees were still forming a continuous crown. The thick bark of the firs had protected their cambium, and their root systems were deep enough that most were buffered from the heat. In some areas, even though the duff had turned to ash and some roots had burned right through, most of the ground was only moderately charred.

Using my graduated trowel to dig down through the layers of the forest floor, I was surprised to find that the soil was wet. It came out in clumps, the silt and clay sticking to my fingers. The aggregates were saturated, the pores filled with liquid and the particles coated in water. It was spring, so it made sense that the soil was still moist from snowmelt, but this felt wetter than usual. I looked up at the canopy and saw gaps in the crowns of the trees. The death of some trees from the flames and the dropping of scorched needles from the survivors meant that evapotranspiration—the combined loss of water through transpiration from the stomata along with evaporation from plant leaves and soil surfaces—had been reduced. With fewer stomata emitting water vapor, the mineral soil in this charred forest remained wetter than it would have been under a full canopy. By the time I reached one meter down, there was a little puddle of water at the bottom of the soil pit.

I took a deep breath. The air was warm, dry, and a little dusty from the ash, and I realized that the water cycling through this forest had been disrupted by the fire. Trees are like geysers, pulling water up from the soil and transpiring the vapor from their stomata. A single mature tree can transpire over 400 liters (105 gallons)—about two and a half bathtubs full—in a single day. Over 40,000 liters (10,500 gallons)—a typical swimming pool—in a single growing season. The forest that Jean, the girls, and I were in today would, under normal conditions, transpire upward of 35 million liters (9.2 million gallons) in a growing season, an amount roughly equivalent to the volume of fourteen Olympic-size swimming pools.

With this amount of water cycling through the trees, I could easily see that the forest was stabilizing the hydrologic cycle in these mountains. The trees were refreshing the air with the water, contributing to rainfall and cooling daytime temperature. When clouds that developed with this moisture deflected the sun, air in the forest cooled even more. The clouds and fog intercepted by the trees also attracted more moisture, causing rainfall to increase to an even greater extent. As the moisture moved up through the canopy, it condensed on leaf-dwelling microbes or volatile organic compounds—such as the monoterpenes that protected the conifer needles from insects and gave them their sharp scent, or the terpenes that helped the trees cool off and cope with heat stress—further enhancing rainfall. When the rainfall hit the ground, the tree roots, mycelium, and soil food web that glued the soil particles together and gave the soil structure allowed even more water to penetrate, thereby improving groundwater recharge. Precipitation thus filtered through the entire forest watershed was delivering pure, clean ground and surface water to the streams and rivers, and ultimately the oceans. Through evapotranspiration, the forest also created a variable air pressure field that transported atmospheric moisture to even greater distances, from one watershed to another. *No wonder it felt so good to be in the forest on a hot summer day.*

I finished sampling, thinking about how soil, water, and carbon were tightly interwoven. Photosynthesis could not happen without transpiration, and vice versa. When canopies are full, the forests regulate the water cycle, and the water cycle ensures that the forests are productive. A productive, intact forest sequesters and stores massive amounts of carbon, and at the same time the carbon in the soils and trees intercepts, filters, and absorbs water. Carbon sequestration and storage are co-benefits of the hydrologic cycle and associated cooling, and these cannot be separated.

"Are these morels, Mum?" Hannah's question drew me out of my reverie. She was pointing to a carpet of plump honeycombed mushrooms poking from the burned duff. The fungal caps, with their black ridges and pale pits, were bursting in abundance among the roots of the trees. Morels are mycorrhizal fungi that form a symbiosis with the roots of Douglas fir, ponderosa pine, or aspen trees, providing mineral nutrients in exchange for photosynthetic energy, but they can also switch to being saprotrophs—decay fungi—to break down charred wood and litter after a burn. Whether acting as a mycorrhiza or a saprotroph, most of the fungus lives in the duff, where its mycelium—long hyphal threads—explore the soil crumbs and pores for nutrient-rich fibers and minerals. Morels fruit in wild abundance after moderate fire, as the ash and water turn the soil alkaline. Pulling a paper bag from her vest, Hannah picked some for our dinner that night. The nutty, earthy mushrooms would be delicious fried in butter.

As we started our hike back, Jean squinted up at the emerald shoots of the live trees against the azure sky and grinned at me.

"Five cents please," she said.

Jean had predicted our trees would survive the fire and she was right. I would have to pay up.

5

GARDENS IN THE FOREST

With the Mother Tree Project seedlings in the ground, our trip to the Cariboo Mountains behind me, and Robyn and Kelly Rose caring for Mum at home, I decided to take a much-needed break. One misty spring morning, I hiked into the forest outside of Vancouver with my friend and colleague Teresa, who had offered to show me the traditional method for harvesting cedar bark to weave into baskets.

Born into the Raven clan of the Gitlan tribe of the Ts'msyen Nation, Teresa had learned this practice from her mother. The two of them gathered cedar bark together every spring in the forests and sold their art at the Heard Museum in Phoenix and at the Sante Fe Indian Market.

"We strip the bark in spring when the skunk cabbage blooms," Teresa said, as we walked past stands of cedar, hemlock, and alder and shallow gullies lush with lady fern and foamflower.

"It's not easy," she added. "When the spring is too dry, the bark can be hard to peel off the tree." Last spring, her mother had tried to pull a strip of bark off a tree that was reluctant to let it go.

"Mom was hanging on to the end of the strip with all her strength and swinging around the tree, trying to get it to release. Then I took a turn, and the top suddenly let go and the piece of bark punched me in the throat!"

Splashing across a narrow stream, we skirted a thicket of salmonberries, and Teresa pointed out the network of bear trails that ran through it. She laughed as I scarfed a handful of the berries, staining my lips a bright fuchsia.

She explained that the first step, here in the southern range of the coastal forest, was to look for a grove of bigleaf maple trees. Cedars and maples grow alongside each other in the moist hollows of the coastal forests, sometimes accompanied by the yews. The trees wick up the ample water with their roots and distribute it through their vascular system to aid in photosynthesis, and ultimately transport the sugars down the phloem just under the pliable bark. The bark of a cedar growing near maple would be supple and healthy, which was important for weaving.

We passed a grove of ancient western redcedars and I raised my eyebrows in enquiry, but Teresa shook her head. The trees were too big, and it would be disrespectful to harvest bark from the elders. We walked deeper into the woods. At last, she stopped in front of a young slender redcedar with a relatively flat surface on its lower trunk. The tall bole was clean of branches, which had been shed by the tree as she expanded her crown and cast shade below.

"Perfect. Nice and straight, not too big, not too small."

Teresa put down her pack. The sun had broken through the mist and the forest was starting to heat up.

"We always ask Sister Cedar for permission to collect her bark," she said. "We explain why we need it."

Laying her hands on the trunk, she closed her eyes and whispered a few words of appreciation and promised to make something beautiful.

If the tree gave permission, it was essential to offer something in return. The weaver who creates the baskets and the forest that provides the bark and roots and other materials have a reciprocal relationship, based on giving and taking in equal measure.

Teresa made an offering of tobacco to show the tree her gratitude for the gift of its bark. Then, using a sharp knife, she made a shallow vertical cut on the trunk. To prevent the harvest from killing the tree, she would take only a small amount of bark, allowing the tree to heal over time.

Using a small hatchet, she pried the bark away from the sapwood and separated the outer bark from the fleshy inner bark, the part used for basketry. Later, she would let the inner bark dry and store

Dr. Teresa (Sm'hayetsk) Ryan in Haida territory, 2017, wearing a traditional hat she wove from cedar bark. Teresa is of the Raven clan, Gitlan house of Xpe Hanax, in the Ts'msyen Nation. She learned to weave from her mother, who, as with many First Nations people, had to learn her traditional craft by visiting museums, where she studied the collections of historic baskets and hats created by her ancestors.

it until she was ready to weave, then soak it before cutting it into strips. Sometimes she interlaced bear grass or canary grass into the cedar strips, making bright symbolic patterns. She also used the bark of birch, maple, and wild cherry and the roots of spruce and cedar in her weaving.

I thanked her for showing her craft to me. Since we'd met four years earlier, she had changed my thinking in so many ways. She'd taught me about the exquisite ability of First Nations people to care for the resources, not overusing or wasting any of nature's gifts, but rather increasing their abundance and health. She had explained how the Ts'msyen followed the laws of nature, maintaining balance through principles of respect, reciprocity, and kinship.

Teresa is a first-generation scholar. As a scientist and an Indigenous woman, she lives and works at the intersection of two disciplines: traditional knowledge and Western science. Her maternal grandfather, a commercial fisherman, had often taken her on his fishing boat when she was a little girl and told her the stories of the salmon. When she turned seventeen, he declared that he wanted her to become a fisheries scientist. She was baffled by this request, but he was adamant. During her master's degree program, she learned the difference between Canadian laws and the laws of the Ts'msyen that her grandfather had taught her about salmon. An Indigenous cosmology is based on things such as the great flows of salmon from the ocean and into the forests with many intricate relations and connections. After getting her PhD, she immersed herself in fisheries science and management, and I had offered her a postdoc position to build her research on salmon and forests. Pacific salmon were teetering on the edge of collapse. Teresa understood her grandfather's vision that she would use her skills to protect the salmon.

Later that summer, we headed to Bella Bella, the community of the Heiltsuk Nation in the central coast of British Columbia. Teresa was

Aerial view of a tidal stone trap, or weir, at Kunsoot River in Heiltsuk territory on the central coast of British Columbia. The half-quadrilateral wall of stones is submerged under the sea at high tide and emerges at the ebb tide. Salmon become trapped behind the wall and are harvested on the lowest ebb tide. The largest salmon are released to spawn, thus maintaining abundant populations. This highly evolved technology, used for over five thousand years along the central and north coast, created wealth and maintained balance in the socioeconomic systems of Indigenous people.

consulting with the nation about researching the ancestral tidal stone-trap fishing technology, which may hold promise for reviving local salmon populations. Long before colonization, the Ts'msyen, Heiltsuk, Tlingit, and Haida nations in the north and the Kwakwaka'wakw and Nuu-chah-nulth nations to the south had constructed and tended hundreds of these salmon traps along the coastline.

We had additional reasons for the visit that day. Teresa had reached out to the Heiltsuk Integrated Resource Management Department about bringing the Mother Tree Project to Heiltsuk territory, and they had invited us to discuss potential collaborative research and take a look at the forests.

At the time, the Mother Tree Project was located across a broad climatic gradient covering the range of Douglas fir in the interior of British Columbia. Teresa, Jean, and I wanted to establish a parallel climate gradient through the cedar-dominated forest along the coast. We had just one coastal site—the University of British Columbia Malcolm Knapp Research Forest, near the city of Maple Ridge, east of Vancouver, and its forests included both Douglas fir and western redcedar. We hoped the Heiltsuk people would be interested in allowing some of their forests to be involved.

Jean, Teresa, and I, along with some of our First Nations partners, had a grand vision of growing the Mother Tree Project beyond British Columbia to become an interdisciplinary international research network. We imagined experimental sites along the entire Pacific coast, a climate gradient that stretched from the cooler, wetter Gulf of Alaska down to warmer, drier Mexico, and even as far south as Patagonia. Our vision included diverse groups of Indigenous and local communities, with people from a myriad of disciplines engaging in common projects, all integrated into this expansive longitudinal study, sharing ideas and data, exploring new questions, working in collaboration to develop and implement a model of regenerative forest stewardship. We knew that such an ambitious endeavor would require more funding, time, and capacity than we currently had available. While it was in the realm of the possible, it remained a dream for now.

WITH PERMISSIONS AND PROTOCOLS from the nation in place, Teresa and I hired a guide to take us into Heiltsuk territory in a herring skiff to explore the forests and identify sites for our projects.

Our guide, Ron, was a Heiltsuk fisherman whose people traced back to the great flood—perhaps a tsunami some fifteen thousand years ago—when they were born into this landscape, as told in their origin story. Decades of hiking the Coast Range, hunting and trapping, and digging for clams along the beaches had given him an intimate knowledge of the territory. Along the way, he pointed out the local Heiltsuk-owned fish processing plant, and he and Teresa talked about the old and new ways of salmon fishing.

It was early August, before spawning season, and there were no salmon or bears yet in the Kunsoot estuary. *Kunsoot,* a word from an ancient dialect of the Heiltsuk language, means "fish traps filled with salmon." The surface of the water was glassy, the treetops veiled in fog. The silence was broken only by the whisper of rolling mist and the high-pitched melody of the resident eagles. Teresa and I grinned at each other, thrilled to be on the water. As we glided over the wall of stones under the tidewater, she described how the people had caught salmon with the passive stone-trap fishing technology when the fish returned to their natal streams to spawn in the fall. Reaching down through the cool water, she brushed her fingers across the algae-covered stones.

"Our people took care of these gifts of creation for thousands of years," she said.

They'd built the stone weirs by piling rocks in long strokes, reaching eight meters in width and a meter in height so the tops exceeded the low-tide mark. The weirs were often hundreds of meters in length, shaped in graceful arcs contoured along the intertidal zone of the river mouths. Each stream, and the salmon population that belonged to it, was cared for by a chief and their community. As they tended the traps, they also created resting ponds in the river for the fish, pulled snags from the channel, and cleaned the spawning beds. The harvest was aligned with the natural cycles of the ocean tides. When the ebb tide receded, the weirs hemmed the salmon behind the stone walls.

In lean years when the salmon populations dipped to particularly low levels, as happens with periodic shifts in the cycles, the return was not touched. Instead, a neighbor would share their traps, or other food was hunted until the stocks recovered. This highly adaptive method not only ensured a genetically diverse salmon population, but helped generate more abundant runs, with larger fish. The ecosystem had remained in balance for thousands of years, and people, bears, eagles, and wolves had plenty to eat.

The giant western redcedars and Sitka spruces lining the coastline had provided the Heiltsuk with homes, clothing, and canoes. In turn, the First Peoples had tended the trees and kept the forest healthy by periodically setting small, low-intensity fires to burn off undergrowth and open up areas for medicine and food plants to grow. Cultural burning had been banned by the Canadian government as part of the Indian Act of 1876, which restricted nearly every aspect of Indigenous life. This byzantine legislation attempted to force assimilation by dismantling their traditional systems of governance and imposing bans on traditional practices, performing dances and songs, or accessing their own territories.[1] Even the fish-trap technology Teresa was taking me to see today had been outlawed by the colonial government during those dark times, and government agents had attempted to destroy thousands of the stone traps and other fishing weirs along the coastline and waterways, as they took the lands, waters, and resources for themselves.

I thought of Mabel Lake in the inland rainforest of British Columbia where my family had horse-logged selectively for generations. My ancestors had been careful to take only the timbers they needed, allowing the forest to rebound from the immense legacy of trees, soil, and seeds left behind. Still, they had not considered the hereditary knowledge of the native Splatsin people who had lived in those valleys since time immemorial and who knew how to care for the resources. As settlers who farmed and logged these lands, my family learned as they went, unaware of the ancient teachings of the people who'd lived there for thousands of years.

Since becoming a forester, I had watched my family's horse-logging livelihood disappear and the big timber companies assume control of the resources, extracting so much wood that the mountains were balding. While the government regulated the harvest, there was a disconnect between economic aspirations and land capacity, and the cut level was set so high that the system was doomed to eventually collapse. The river drives around Mabel Lake became so congested that logs choked the Shuswap River. The Splatsin people who had depended on and cared for the salmon for millennia were pushed aside and the spawning beds became clogged with logging debris.

Ron hopped out of the skiff with barely a splash and steadied the boat for us to follow. Teresa jumped back in the nick of time as I landed feetfirst in the shifting bank of pebbles and clamshells, soaking myself from head to toe. She teased me for being so clumsy and I gave her my best stink eye. Wiping down my plant book, I shook the water out of my gum boots. I felt solace and ease here in the Pacific coastal rainforest, my worries about the many wildfires now burning in the interior of the British Columbia temporarily out of mind.

My cruiser vest was wet, but my camera and laser hypsometer—an instrument that measures the heights of trees—were tucked in a pocket and had stayed dry. A flock of sandpipers ran by on the mudflats, their long legs moving faster than my eyes could follow. "*Wheet-wheet-wheet*," they whistled. Every few meters, they stopped to probe the mud for crustaceans and worms, their long, sensitive bills exquisitely evolved for survival in this ecosystem. An eagle nesting in the branches of a yellow cedar called "*Ki-ki-ki-ki-ki-ki-ker,*" and her mate flew in with a catch for their eaglets.

"*Prruk-prruk-prruk,*" a raven gurgled, and Teresa returned his call with a "*Toc-toc-toc,*" revealing her kinship to the shimmering ebony bird.

Ron pointed out a Heiltsuk smokehouse at the edge of the forest. Once the salmon were hauled from the fish traps onto the shore, the fish would have been cut, dried, and smoked for the winter. The salmon bones were returned to the water with gratitude and respect, to fortify the water with nutrients like calcium and magnesium.

Teresa trained her binoculars on a rock wall at the lowest intertidal zone as our boat slowed into the bay. We were anxious to see a stone trap.

"That looks like a clam garden." She pointed out how each rock had been carefully rolled into place to create a terrace, an ideal habitat for clams to reproduce along the shoreline. The sediment that accumulated behind the rocks enabled the butter clams and long-necked

Logs choking the Shuswap River en route to the A. R. Rogers Sawmill, Enderby, in Splatsin te Secwépemc territory, ca. 1950. The Splatsin people who relied on this river for salmon were forced onto tiny reserves while white settlers were given their land to homestead. The river mouth at Mabel Lake where my grandparents lived was thirty-five kilometers upstream. Once the large timber companies were granted licenses to cut, small operators like my grandparents were forced out of the trade and excessive log drives would choke the river.

clams to flourish. The clam garden was designed to be underwater at high tide and exposed for harvesting at low tide. People had shelled clams here for hundreds of generations.

I realized that the pile of clamshells I was standing on could be a shell midden, an ancient kitchen compost containing the shells from thousands of meals. Archaeologists had found clam middens dating back about six thousand years at the nearby village site of Húy'at.[2] I dug my hands into the mixture of broken bivalves and moist soil. This earth was as dark as black prairie soil and probably richer.

Teresa and I scrambled up the bank, wading through layers of pink springbank clovers, yellow-flowered silverweed, and chocolate-petaled riceroot.

She knelt among the luxuriant blooms, excited to see the old terraces.

"These are root gardens." She explained how they had been constructed and tended over generations to cultivate starchy foods to supplement the protein-rich marine diets. "The roots are delicious with roasted clams and fish eggs."

My stomach rumbled at the thought. I was reminded of the rice terraces Jean and I had seen decades earlier as university students backpacking our way through China. It struck me that the tilling practices in the watery rice terraces were probably informed by age-old cultivation techniques like these.

Along the sandy shoreline hemmed with Sitka spruce, Teresa explained that these were ideal places to gather spruce roots for weaving baskets and hats. She described how to carefully dig the gleaming white radicles out of the sandy soil.

The towering forest edge looked far away. As we approached the bramble of salmonberry, I spotted a crabapple tree.

"There's a plum tree over here!" Teresa called. Noticing the distance between the fruit trees, I realized that these, too, had been tended. The salmonberry bushes had been cared for as well, thinned and pruned. The black huckleberry, thimbleberry, and red elderberry were thriving and plump with fruit. The entrails from the fish had been used to fertilize the berries.

"I'd always been taught that coastal First Nations people are hunter-gatherers," I ventured, though it was quickly dawning on me how erroneous this belief was.

Teresa rolled her eyes.

"That is such a misconception. We cultivated these foods long before Europeans got here." She spread her arms to take in the river, the berries, the fruit trees, all the edible riches surrounding us.

"We had highly productive food systems. We had our own food gardens. We used cultivation practices that were refined over thousands of years. We had *wealth*."

She had every right to be frustrated. The myth that Indigenous people had lived a subsistence existence before colonization was deeply embedded in the colonial imagination.

"Our culture, our connections to the land, our sustainable food systems—all of it was far more advanced than we are given credit for."[3]

We stood and looked over the estuary, the water sparkling in the high noon sun. It was teeming with life, and the soils were as rich as any I had seen. In this place of abundance, First Nations families had stewarded fish and shellfish populations, developed sophisticated passive fishing technologies, built and nurtured root, berry, and clam gardens, and helped shape monumental cedars to provide them with poles and canoes. These highly refined agrarian and marine technologies had ensured an ample and steady supply of food while keeping the land healthy and productive for future generations—an achievement few contemporary Western cultures had been able to match.

A MONTH LATER, during salmon spawning season at Nimmo Bay, in the southern reaches of the Great Bear Rainforest, I gained a more visceral understanding of the profound connection between salmon, bears, and trees.

For the past two Septembers, I'd helped Irvin Speck, a member of the Gwawaenuk Tribe of the Kwakwaka'wakw Nation, lead visitor walks through the forest to an ancient cedar mother tree. This elder tree had been here long before settlers had arrived in British Columbia. The evening before the walk, Irv and I traveled by boat to an old village site at the northwest end of the bay. After a century of dislocation, the Gwawaenuk people were returning to their land to resume their rights to tend their resources. Irv was doing his part, living in his people's traditional ways.

As the evening sun sank behind the windswept cedars and hemlocks on the ridgeline, he pulled on tall gum boots over faded jeans. A sea lion poked his whiskery head out of the midnight-blue water as our skiff headed into the bay. Irv's face was peaceful, and he looked at home in his territory, yet I knew the clearcut scars visible on the nearby hills caused him pain, knowing the effect it would have on the next generations.

We moored on the rocky point and picked our way across the barnacle-encrusted stones and urchin-filled tidepools along the shoreline. Irv immediately spotted a grizzly matriarch upending boulders in search of crabs. Swinging her massive head from side to side, she toppled one boulder after another, revealing the formidable strength in her paws and the precision of her claws. I felt adrenaline surge through me as I searched the pockets of my vest for my canister of pepper spray. Irv chuckled, then picked up two rocks and banged them together, communicating our presence to the bear. She glanced at us, then continued undeterred in her quest for dinner. When we were about a hundred meters away from her, we ducked onto the path heading into the forest.

I followed Irv through prickly salmonberry, the broad leaves swishing against my arms. We stooped to clear the trail of encroaching canes and fallen branches as we approached the base of the giant cedar mother tree. I peeked inside the hollow at her fluted base to see whether the den had been used by a mother bear and her newborn cubs last winter. I noticed a length of barbed wire stretched across one lane of the trail as we turned toward the river. Irv explained that the wire would scrape hairs off of passing bears, which were then collected and sent to the lab for DNA analysis to trace the genetic lineages of the bears returning to the river.

We headed toward the river, the sound of the rushing water growing louder and the watery air breezier with each step. Within minutes, our trail was meandering among sandy pools and mossy hummocks shaped by the ever-changing currents.

"Look at this!" I turned in a half circle, gazing around. No matter how many times I'd seen this sight, it always filled me with awe. The forest floor was carpeted with partially eaten salmon carcasses. Most of the fish just had heads remaining after the bears had gorged themselves on the protein-rich roe in their bellies. The trail was slick with purplish oily scat, the salmon protein mixed with salal and salmonberries. In some places, we could see the vertebrae of the previous year's run, with pairs of opal pin bones clasped like hands. The bears carry their catch into the forest so they can eat in peace, avoiding clashes with other bears. The leftovers provide a feast for the red foxes, martens, ravens, bald eagles, and smaller birds and insects, who scavenge the remains after the bears have eaten their fill.

Studies would show that salmon density during spawning season can average five salmon per meter and that a single grizzly could carry 150 fish a day into the forest. From what we could see at Nimmo Bay, those counts were easily reached here. The bear scat, urine, and salmon tissue broadcast across the forest floor would decay over the winter and seep into the soil, enriching it with a flow of ocean-derived nutrients. All that would remain the following spring would be the bones of the decayed carcasses.

Above the roar of the river, we heard the sound of splashing and branches snapping. Through the trees, I could see the elderberry bushes lining the river swaying. My face turned hot and my chest started to thump with fear. Without waiting to see if the movement in the bushes was caused by a bear or the current, we quickly turned and headed back toward the boat. Irv moved so swiftly through the trees that I could barely keep up. At the shoreline, we scoured the estuary, but the grizzly matriarch was nowhere to be seen. As we motored out of the bay, I looked at the clearcuts in the surrounding mountains and knew that the bears were in far more jeopardy than we were.

The following spring, I visited my doctoral student Allen at his humble grad student desk down the hall from my office at UBC. Allen, Teresa, and our crew of technicians had been making regular trips to Bella Bella for several years to examine the effects of salmon density on riparian forest soils. He had just defended his dissertation, and we were meeting to discuss the papers he would be publishing.

Allen's findings showed that increasing salmon density had broad impacts on riparian soil fertility.[4] He had quantified the effects of the decaying salmon on soil nutrients and used the latest metagenomics techniques to describe the fungal and bacterial communities. At streamside ecosystems with greater numbers of spawning salmon, the soil was richer in nitrogenous compounds, phosphorus and calcium macronutrients, and manganese and aluminum micronutrients. The greater soil fertility was accompanied by shifts in the mycorrhizal fungal and bacterial communities, with certain groups of species flourishing in nitrogen-rich soils. These microbes were well adapted to decomposing the highly proteinaceous tissues. At rivers that had greater salmon-derived nitrogen, Allen also observed that the fruiting mushrooms were species that preferred soils replete in nitrogen. Using the heavy isotope ^{15}N as a tracer, Allen found that these mushrooms also contained signatures of salmon nitrogen in their dense flesh.

What had me most excited, though, was his estimate of the carbon stocks in the salmon-rich forests. I nearly leaped out of my chair when I read his first chapter reporting that the total amount of carbon in the soil was about six hundred metric tons per hectare—considerably greater than other ecosystems studied in the region.

"These soils are off the charts!"

Allen scratched his beard. "It's kind of an incidental finding."

I knew that quantifying carbon pools wasn't the main objective of Allen's work. Yet here was this mind-blowing estimate, staring us right in the face. The carbon stocks in salmon soils were three times those of the wettest forests in the Mother Tree Project at Malcolm Knapp, which occurred in a submaritime landscape on the Pacific coast. They were *six times* that of the driest forest in our climate gradient, Jaffray, located in southeast British Columbia, where the soils were not directly shaped by salmon. Along with the climate and inherent soil characteristics, the greater nitrogen availability in the salmon-rich forests led to greater tree and plant growth, and this contributed to greater biomass and carbon accumulation in the soil.

The luxuriant forests in the midcoast region repay the salmon by shading and cooling the rivers. The forests also return nutrients to the waters and provide downed wood that helps create habitat for wildlife. Old trees that fall into the channels create deep pools where the salmon can rest. The new fry that emerge in the spring thrive in the shaded organic and nutrient-rich pools, where they feed and grow before migrating out to sea. The returning salmon enrich the people, bears, and forest, closing the circle.

I felt exhilarated just thinking about the primordial cycles of renewal and regeneration. If we could work under the guidance of the Heiltsuk people in establishing a Mother Tree Project site in their territory, we could learn far more about these productive and remarkable ecosystems, whose forests were enriched by the salmon in the soil, the legacy of ancestral stewardship, and the productive coastal climate. Teresa and I had also been discussing new collaborative research to understand how salmon nutrients influenced the cedar trees, and how the nutrients might travel from tree to tree through mycorrhizal fungal networks.

I thought back to a research trip to Bella Bella with Allen and Teresa in 2016. I'd just done my first TED Talk and was still glowing from the heartfelt reception to my work. My brother-in-law, Bill, had spent several days helping me draft the script and practice the talk so that it followed an engaging storyline, and it had caught people's attention. The public had embraced the concept of the mother tree with more enthusiasm than I'd ever imagined, proving that people hungered to reconnect to nature. Just a year earlier, I'd completed chemotherapy and was still healing in body and mind. *How things change.*

I was grateful that cancer was in the rearview mirror and I would be sticking around to see my daughters into adulthood. At the same time, the experience had had a lasting effect on me. My battle with breast cancer had changed my approach to life by changing my relationship with death. Being forced to accept the limits of my own existence had driven home to me, as nothing had before, that we are all part of an interdependent web of not only other people but every breathing and nonbreathing bit of creation. Witnessing how the natural world embraced the inevitable had helped to ease my fears of my own aging and death. In the salmon forests of British Columbia, we'd uncovered yet more evidence of the rich dance of life and death, of how death creates the currency for the next generation. The forest traps carbon and cycles water, and the salmon return, not only to rivers but also to the land, to reside in trees that are a thousand years old and could live another thousand. The interconnectedness—salmon living in the trees for centuries—was reality.

Year after year, I had watched the forests decline with broad-scale industrial logging and I'd seen the vibrancy of the regenerating stands pushed further and further out of balance. The replacement of old forests with monoculture plantations and the extinguishment of native plants with pesticides was making the forests sick, yet the commodification of our landscape continued relentlessly.

Teresa had shown me that a different way was within reach. At Kunsoot, I had seen with my own eyes how ecosystems could be tended for benefit of both the forest and people's livelihoods using sophisticated technologies that dated back millennia. I had understood a small form of this land ethic from my grandparents, but had not seen it reflected in contemporary industrial management. Instead, colonial management systems were leading us from abundance to scarcity.

But the salmon story was full of hope. It showed how people with ancestral knowledge systems centered on tending the Earth according to the natural cycles and connections had intentionally protected and enhanced the resources where they lived. The gardens in the forest confirmed that humans were part of nature, not separate from it, and that the land itself was ingenious, resilient, and hardwired for health. We had the opportunity to rebuild from this ancient wisdom and resiliency to regain balance. To become guardians of our environment instead of exploiters, to give back at least as much as we take—like the weaver and the cedar tree.

6

THE MIRACLE OF YEW

"My brain is a fog," Mum said, waving her hand over her forehead. "I can't think anymore."

We were sitting at her kitchen table, its surface hidden under piles of unpaid bills, empty envelopes, and mounds of folders in a rainbow of colors. Mum used color-coded files as a strategy to keep organized, but now these looked like unsorted jigsaw puzzle pieces.

"It's past time, Suzie," she said. "I want to apply for MAID. I need your help."

MAID was the acronym for Canada's medical assistance in dying program. This wasn't the first time Mum had brought it up, but we'd never really talked seriously about it. Before I could form a response, she plunged ahead.

"I've been thinking about this. If I wait too long, it'll be too late."

"But Mum, there's still time—"

"No, Suzie. Pretty soon . . ." She didn't finish her sentence, but I knew what she was thinking. Every day, it seemed, she lost more of her ability to remember things, grew more discouraged. I knew she was deeply frightened at the prospect of losing all her memories, her identity, the core of her being. Mum loved life and she wanted to live, but not at any cost. She had watched her own mother, my Granny Winnie, linger with Alzheimer's. MAID could give Mum a way to preserve her dignity, to exercise control over what future was left to her. This decision was hers and she was sticking to it. Once Mum made up her mind to do something, it was a done deal. She had a stubborn streak as wide as the Columbia River—which she'd once swum across on her own, just to prove she could.

I thought back to a decade ago, when she had been forced to give up her driver's license on the recommendation of her doctor. Losing her license had been a huge blow. Her greatest joy was the independence she got from driving her silver Camaro anywhere she wanted. She relished her freedom, needed to be able to flee over the mountains and along her home lakes of the Kootenays. Without her car, her wings had been clipped. Though we had tried to talk her out of it, Mum had decided she would try to get her license back.

For weeks, she had applied herself to the multiple-choice knowledge test, studying day and night, and then she took the written exam and got her learner's permit. Then came the driving lessons. Robyn, Bill, and I helped her practice while at the same time teaching our own teenage daughters how to drive. Being in a car with Mum at the wheel was terrifying. She would sail through four-way stops without pausing, giving us hell when we tensed up. She took hard rights in front of cement trucks. If she forgot where she was going, she'd just start following the car in front of her. She would scold us when we suggested she turn off her unneeded blinker. She took the road test, shushing the examiner—while he was giving her instructions—so she could focus. She failed the test the first time, then insisted on trying again. Hannah, Nava, and Kelly Rose all passed their tests while Mum failed a second time. After the third time, the licensing bureau told us to stop bringing her back. It was a humbling moment for her. Mum, who had raised three kids as a single mother with a full-time career, outdone by her teenage granddaughters.

But she gathered herself back up. She took up walking, sometimes heading downtown twice in one day, often covering twenty kilometers. She heaved rocks in her backyard, weeded her flower beds, planted a pear tree. She redesigned her garden to require less weeding and watering, not only because it was less work, but because she worried about drought and the community water supply.

But now, she couldn't remember how to pay her bills or which button to push to turn on the television. She'd had to quit skiing because she was afraid of falling in a tree well or being unable to get off the chairlift. Even worse, she couldn't read her favorite books or properly care for her plants.

"I am too tired and confused to look after the garden," she said. "I've done my best, Suzie. It's time for me to go."

Reluctantly, I agreed to look into MAID, just to put her at ease. Mum immediately cheered up because she felt more in control of her life. But I dreaded the thought of losing her.

AFTER MAKING MUM coffee and toast, I walked home, where Amanda and Eva were waiting to celebrate the publication of Amanda's first paper. They'd brought a six-pack of Amanda's favorite ale and a bag of chips, and Eva had brought her new border collie pup, Tia. For as long as I'd known her, Eva had been dreaming of getting a dog. But a graduate student's life is anything but stable, so she'd made herself wait until the conditions were right. Now she and Tia were inseparable.

Amanda had been keeping a grueling schedule. Just back from playing baseball in Japan, she was heading to Golden to collect more soil carbon data before the snow flew. The ground was almost frozen and the days were getting short.

"I'm really proud of you, Amanda. It's not easy juggling all this," I said, as I poured the midday beer into three tall glasses.

"Oh, it's nothing," she said, her cheeks dimpling. Her walnut-brown hair was tied up in a bun, nothing fancy, but ready for action.

"I have it pretty easy," she added. "Other people are doing much more than me."

I hid a smile. I was pretty sure this wasn't true. I poked her.

"I don't know anyone else who got their master's and their PhD while playing baseball in the world championships," I pointed out.

"Yeah! Cheers, Amanda!" Eva said. From her place at Eva's feet, Tia looked up to see what all the fuss was about.

We held up our glasses, swigged our beers.

I thought back to celebrating my own first discovery with my doctoral adviser, David Perry. After finding that carbon-13 and carbon-14 transmitted between birch and fir, we'd shared a glass of whiskey in his office. I recalled how excited we'd been at the finding that the two species were trading photosynthetic carbon back and forth in a reciprocal relationship, and that a greater amount of carbon flowed from nutrient-rich paper birch to the more acidic Douglas fir, especially when the fir was shaded. This was the first sign that the trees were in direct communication through underground channels, forming a robust, interdependent network.

Despite her humility, I knew Amanda was pleased to be celebrating this moment. It had taken her a decade to complete her graduate studies—from her twentieth year to her thirtieth—and we were finally toasting her first dispatches to the world. In her new publication, Amanda reported her clearest, most profound finding: interior Douglas fir seedlings were larger and had more foliage when growing in the neighborhood of kin Douglas fir seedlings rather than strangers.[1] It didn't matter if those strangers were other Douglas firs, or some other species entirely.

This finding alone was astounding. It meant that trees recognized other trees that were their relatives and benefited from growing near them. This complemented Amanda's earlier master's research. She'd found that seedlings establishing next to older kin rather than strangers not only had more productive traits, but greater mycorrhizal colonization rates, presumably because they had access to the established mycorrhizal network of the older sibling.

The older trees, with more resources, were connecting with and nurturing their younger siblings.

These discoveries were breathtaking. Not only did they fly in the face of modern forestry practices, but they corresponded with thousands of years of Indigenous knowledge on the importance of kinship among living beings.

Over the previous three decades, a common forestry practice in the dry Douglas fir forests had been clearcutting followed by planting. This was done to generate more profit and reduce planting costs, speed up rotations, and reduce uncertainty compared with selective harvests reliant on natural generation. That's why the companies were clearcutting the Douglas fir forests in Nlaka'pamux territory, Dave Walkem's ancestral home. These tree farming practices meant the next forest could be cut faster and more money could be made sooner.

Getting rid of the old trees, they believed, also reduced risk of wildfire and the spread of bark beetles. The mother trees were thus sent to the mills, and new seedlots of seedlings were planted to replace them. But these planted trees fared poorly in the arid environments, often succumbing to heat, drought, or frost. Nursery-grown trees were bred for rapid growth and straight trunks, rather than belowground traits or adaptive cues enhanced in the presence of relatives that could increase survival. Perhaps because of this, their rooting responses were not finely attuned enough to cope with extreme drought in June, or their needles were too sparse to withstand a sudden frost in July. Natural regeneration around parent trees, with their locally adapted seeds, adaptive cues to relatives, and guerrilla root systems, was proving more successful.

But there was even more to this story. Amanda wanted to know how kin selection played out in more diverse forests, such as those in Prince George, where she'd grown up. There, she'd seen Douglas fir growing alongside lodgepole pine, spruce, and aspen. Could a tree's neighbors—their species, density, or even the soil in which they grew—affect how well Douglas fir related to its own kin?

As pitcher and captain of her team, Amanda knew that camaraderie affected the performance of all the players. Just like she did in our lab, she would boost the morale of the group, and this increased cooperation while working together on one another's projects in the woods or in the greenhouse. Her natural intuition led her to ask questions about how the whole forest influenced kin relationships.

Amanda designed her experiments to look deeply into what diversity in the forest meant to relationships among kindred trees. She planted Douglas fir seedlings in mixed neighborhoods of kin seedlings, stranger seedlings, and a different species altogether, lodgepole pine. She found that the trees were able to adjust their behaviors based on the complexity of their quarters. Even in mixed stands, kin seedlings preferred to grow near kin, but they performed better in the neighborhood of lodgepole pine than around other Douglas firs that were unrelated to them. They responded to neighbor identity by adjusting their mycorrhizal status, slenderness, foliage, and fine root allocation. In these mixed stands, the kin seedlings were able to integrate their complex environments and respond in different ways to enhance their performance.

This made sense—interactions between kin depended on who their neighbors were. In this case, the pines were smaller than the firs and less competitive, and this enabled kin to thrive in their siblings' presence.[2]

Amanda turned to Eva and nudged her. "We'll be celebrating *your* experiments next year."

"Eek!" Eva pushed her glasses up the bridge of her nose and buried her fingers in her dog's soft fur. I knew she was awed by Amanda's accomplishments. Amanda was her mentor, her beacon of hope, having graduated when Eva started her own doctoral studies.

"You can do it," Amanda said. "I'll help."

In one of Eva's doctoral experiments, which she'd nicknamed the Septuplets for its planting pattern of seven trees in each treatment, she aimed to investigate how the transmission of photosynthate among four species of trees affected their relationships and interactions.

In the inland rainforests of British Columbia, cedar towers in the overstory, with Douglas maple growing in the midcanopy gaps, and the little Pacific yew rising up beneath. These three species tend to grow near one another due to their ability to layer in generation after generation of saplings through vegetative reproduction. Like cedar, the drooping branches of yews take root and form new trunks where they touch the ground. Maple, too, returns in layers, or sprouts from basal buds, and a smattering of new genetic variation comes from the seeds of all three species.

The yews in particular are dependent on the cool, moist habitat of the old forest and are especially prevalent in gullies and along streams, where their sinuous stems and swooping branches are coated in communities of moss. Their emerald crowns shade the creeks and their wine-red roots stabilize the banks, creating bassinets for the spring hatch of salmonids and other fish species. The red fruity arils coating the yew seeds are eaten by birds, rodents, and deer, though its bark and leaves are poisonous—the tree's scientific name, *Taxus,* is Latin for "toxic." The craggy evergreen tree has long symbolized death and resurrection. In medieval Britain, yews were planted in churchyards and cemeteries, and yew boughs were tucked into coffins to accompany the dead into the afterlife. In Shakespeare's *Macbeth,* the three witches brewed a potion that, along with deadly hemlock, included "slips of yew, silver'd in the moon's eclipse."

Cedar and maple make their homes among even taller citizens of this forest, such as the rangy ectomycorrhizal hemlocks, seeding on rotten logs and into the shadowy forest floor, and the Douglas firs, larches, and pines, each regenerating from seed in the larger canopy gaps. Eva had already figured out, using molecular genetic analysis, that cedar, maple, and yew were colonized by almost a dozen of the same arbuscular mycorrhizal fungal species. We figured that the three companion trees were linked in a fungal network.[3] In the rainforest, tapping into the arbuscular mycorrhizas of other hosts increases the odds of a seedling surviving. As the seedlings germinate in the shade of other trees, they can take advantage of the large pool of dissolved nutrients already available in the established network. This could enable the seedlings to survive, even as older trees are usurping resources—light, water, and nutrients—and perhaps hosting infectious pathogens.

The arbuscular mycorrhizas of cedar, maple, and yew are distinct and exclusive from the ectomycorrhizal fungal networks of overstory firs, hemlocks, and pines. The ectomycorrhizal tree species live intimately among the three arbuscular sisters in the rainforests, but their fungal communities, and hence their connective webs, are separate. Arbuscular mycorrhizal fungi, which colonized the roots of the first land plants about 460 million years ago, live inside the cortical root cells of the host plants, whereas ectomycorrhizal fungi, evolutionarily divergent from arbuscular mycorrhizas about 50 million years ago, remain on the outside of the plant cell. The larger Douglas firs, hemlocks, and pines have high fidelity to ectomycorrhizal fungal species and do not associate with arbuscular mycorrhizal fungi, and—vice versa—the cedars, maples, and yews do not form ectomycorrhizas.

Similar to my own doctoral research three decades earlier, Eva wanted to test whether photosynthetic sugar moved underground from the overstory cedars to the understory yews and their companion maples, and how this compared to potential fluxes through the soil into incompatible ectomycorrhizal Douglas fir.

THE YEW TREE in our forests was significant to Eva and to me for other reasons, too. In its cambial tissue, the yew produces a secondary defense metabolite, called taxane, for protecting itself against pathogenic fungal infections. Among the taxanes is the anticarcinogenic compound paclitaxel.

In 2012, when I was diagnosed with metastatic breast cancer, I was treated with paclitaxel. The medicine exhausted me as it worked to block the growth of my cancer cells. My chances of recovery, which were 50 percent with surgery, rose to 80 percent with intravenous infusions of paclitaxel along with a concoction of other medicines. As the winter-bare trees outside my hospital window bent toward one another in communion, I welcomed the medicine gliding into my blood. After months of chemotherapy and radiation, the last drop was streamed into my body. I had been well for over a decade since. Paclitaxel helped cure my father from lymphoma the year after my own cancer treatment, and now it had become a salve for Eva's mother, who was herself battling breast cancer.

The First Peoples who lived in these forests had long known of the yew's medicinal properties. They brewed teas and made salves from the tree's bark, needles, and berries, treating everything from skin diseases to neurological disorders. In the Haida language, the yew was called the "bow plant" for its incredibly strong wood. In Kwakwaka'wakw culture, the tree was the "chief of the plant spirits." The Nlaka'pamux used parts of the yew in ceremonies for those who had passed to the spirit world.

The pharmaceutical industry had stumbled upon yew's anticancer qualities in 1962. Paclitaxel was found to be curative for metastatic ovarian and breast cancers, and later, esophagus, lung, and many other cancers. This had sparked a gold rush for yew bark in the 1980s and '90s. Scientists learned to extract paclitaxel from the bark of Pacific yew, but it took 2.3 kilograms (5 pounds) of paclitaxel—or 27,500 kilograms (60,000 pounds) of yew bark—to treat a single breast cancer patient. Poachers hunted the yew in the misty rainforests of Canada with the intensity of ivory poachers. Hundreds of thousands of the trees were destroyed each year.

In my wanderings through old-growth forests back in the 1980s, every yew I found had been violently stripped of its purplish bark, its naked sapwood exposed. Without a shred of cambium remaining, the trees were left to die. I held out hope that the fungal endophyte *Taxomyces andreanae*, later isolated from Pacific yew's xylem tissue and found to biosynthesize paclitaxel, would save the tree species. But the fungus produced the medicine only in very small amounts, and could not be substituted for the precious cambium tissue.

The yew tree was also being demolished by the logging industry. When the overstory was clearcut, the ancient parents and offspring shriveled in the absence of their lifelong companions. Any trees or rootstocks that survived the feller bunchers were unable to recover or reproduce in the open, and ultimately perished. The mystical yew, revered in folklore and medicine, was under threat.

But complete extinction of the Pacific yew was averted in the nick of time. Scientists learned to make the modern pharmaceutical Taxol in the laboratory using semisynthetic cell culture. Different species of yew with more productive cell lines were discovered and used for tissue culture.[4] Unlike the Pacific yew, three of these species could easily be cultivated in commercial nurseries and are so common they can even be found planted in hedgerows and gardens in towns and cities. Today there is ample supply of the medicine.

I recalled Teresa gathering cedar bark for her baskets, never taking more than she needed or what the forest could provide. I remembered the gnarled old yew that grew in the woods near Grampa Henry and Granny Martha's wooden house at Mabel Lake. Grampa Henry used to adorn the little tree with lights when the snow fell, celebrating and honoring the spirit of the forest.

After my last round of paclitaxel infusions, I had brought Hannah and Nava to the forest to introduce them to the lifesaving trees. The three of us had danced through the yew grove, rejoicing, and I had promised my daughters that one day I would help protect this divine tree and honor its medicine. In designing the Septuplets study, Eva and I had used our scientific skills to ask the forest how we could show our gratitude for the yew and reciprocate its gift of healing.

SETTING HER GLASS DOWN, Eva took her notebook from her satchel and sketched a figure that looked like a wagon wheel, with six spokes radiating from a central hub. Amanda leaned in for a closer look.

"At the hub of the wheel I'll plant a cedar tree—that's the mother tree. At the end of three spokes on one side are a yew, a maple, and a Douglas fir, and there's another set of yew, maple, and Douglas fir on the other three spokes."

One set of trees—one half of the wagon wheel—would be left in ambient light, while the other set would be covered in shade.[5] She would plant six replicates of the Septuplets at each of the six Mother Tree Project sites, for a total of thirty-six in the entire experiment.

"What I'd like to find out," Eva said, stroking Tia's ears, "is what influences the yew's ability to produce paclitaxel. Is it the overstory trees, or the light it receives, or the soil, or the overall health of the forest? What has the most influence on the quality of the medicine or how much the tree produces?"

It seemed to me that the fungal trellis shared among the cedars, maples, and yews could be essential in the establishment of new saplings, or in the mediating of relationships and energy or infochemical flows among the trees, and thus the exuberance of the ancient tree in the understory. If the cedar cast more shade on the yew, perhaps it helped supplement its energy by shuttling more photosynthate through its mycorrhizal fungal connections, possibly influencing the quality and amount of medicine produced. The cohesiveness and composition of the whole forest, and how this influenced the vitality of individual trees, could be important to the tree's medicinal properties. Might the medicine be more powerful in a lovingly tended forest where the trees were protected or shaped or fed?

Excitedly, we posited more questions to each other. If carbon moved belowground from tree to tree, would it travel through the arbuscular mycorrhizal fungal network? It was well-known that photosynthate flows from leaves to roots to arbuscular fungi. But would the photosynthetic carbon then move through the arbuscular mycorrhizal fungal highway and into neighboring plants? Or would it be exuded by the mycorrhizal hyphae and meander through the soil to nearby roots?

Regardless of the transmission pathway, the big question in Eva's mind was, Would the yews benefit from the overstory cedars and maples, boosting their survivorship, defense chemistry, and thus the diversity of the forest?

If ectomycorrhizal Douglas fir received nothing from cedar, while the arbuscular mycorrhizal neighbors were enriched, this could provide evidence that photosynthetic carbon indeed moved from cedar to yew or maple through arbuscular networks. Any carbon received by Douglas fir would have to come through the soil, since it was not connected to the others.

If shade was cast upon yew or maple, as happens in old forests, Amanda noted, would these species receive more carbon from cedar?

"The more shadowy the old forest, perhaps the more carbon would be shared to support the diverse understory," she said.

Diversity was increasingly known to boost productivity in the entire forest, reduce damage by disturbances, and enhance resilience. Would greater survivorship of yew in the old forest also provide food for birds and mammals, enhancing the functioning of the entire food web?

If that was the case, Eva said, could those carbon subsidies also augment the production of paclitaxel?

When the Septuplets seedlings were big enough, Eva would add carbon dioxide tagged with the carbon-13 isotope to gas-tight bags covering the trees, a process called isotopic labeling, allowing her to trace the carbon fluxes from tree to tree. Then we might find the answers to our questions.

"Even if the only thing that comes of this study is a different way to describe and talk about and think about these trees and their relationships in the forest—that would be so, so valuable," Eva said.

She drained her glass and set it down on the table.

"I've been thinking," she said. Amanda and I sat up straighter. You never knew what direction Eva's thoughts would take, but these three words always prefaced something interesting.

"Everything that we're getting at in our lab, all of these ideas about connection and respect and reciprocity in the forest—these are very, very old ideas. Like, these are not new. These are not *our* ideas. The original people of this land had it right. And we're sort of bumbling along trying to catch up. Western science is like this *baby*."

I smiled at the analogy, but she was spot-on. I remembered Dave Perry, who was a descendent of the Athabascan people, telling me, "We believe the big trees nurture the little trees."

Eva ate a few chips, then continued.

"Just the way that thought has developed in Western science is like, you need to have one little piece, and then another little piece, and you need to study these little pieces really, really well—but there isn't much bigger picture thinking. You study to see what this root does and what the other root does, and then you make a hypothesis. But there's no guiding moral compass. There's objective truth—but there's also traditional ecological knowledge truth, which is also a science, and it's truth. It's definitely not like Western science. There's the truth that is Western science, and then there's the truth that is wisdom. We should be asking, What is the wise interpretation of this? What is the wise way forward?"

Eva stopped, suddenly looking self-conscious, but Amanda and I were both nodding in agreement.

I always reminded my students not to get lost in the details, to look at the big picture. That there was always more in a forest than what you could see with your own eyes. Eva had taken this teaching to heart. I was so proud of these young women, so impressed by their intellects and integrity. Someday in the future, Jean and I would pass the torch to students like Eva and Amanda, and I had every confidence that they would carry on the critical work of speaking for the forests. Because Eva had just asked the question every scientist, every leader, every human being should be asking right now.

What is the wise way forward?

7

SISTER CEDARS

I held my breath as the turboprop plane bumped over the Haida Gwaii archipelago. Islets of beryl forests winked in and out of the rain clouds. Jean, Teresa, and I leaned forward in our tiny window seats, enraptured by the chain of islands below us—the ancestral home of the Haida Nation. These 150 volcanic islands, formed when huge tectonic plates moved under the Pacific, were separated from mainland British Columbia by Hecate Strait, a shallow passage known for its ferocious waves and merciless winds.

The Pacific westerly weather carrying us to this oceanic string of pearls brought rivers of rain and mist. From this maritime climate grew one of the most productive rainforests in North America, with trees towering sixty meters in height, four meters in girth, and up to two thousand years of age. Sometimes called Canada's Galápagos, the archipelago was a haven for biodiversity.

In the seat next to me, Jean leaned forward in anticipation. She had come to know these forests as a summer student thirty-five years ago, when she'd worked for a logging company at a camp accessible only by boat or air. Her job was to survey the small seedlings the company planted after stripping the watersheds of their ancient forests. She'd told me about the colossal, otherworldly trees she'd seen, the Sitka spruce crowns that spread across whole estuaries, mosses that grew as deep as ponds, northern goshawks that preyed on entire forest food webs, rivers so packed with salmon that their bodies almost spanned the banks. These forests were in rhythm with the land, sea, and people. In our twenties, she'd regaled me with story after story on our many hiking and biking trips.

Jean was the only single girl living at the logging camp. She had a boyfriend who kept the other guys at bay, and she'd fallen in love—most deeply with the islands, the jeweled coves and rocky cliffs, the long sandy beaches, the mists swirling around centuries-old ancestral cedar poles. The company logged day after day that summer, filling freighters that would ship most of the massive logs to the mainland for processing. Jean loved her work, but this felt like a betrayal. No one had asked the Haida if the loggers could take their forests. The people who'd lived here for fourteen thousand years had not been consulted. Haida law, upheld by hereditary lineages that tended the resources, was ignored.

Thousands of great trees were felled.

In 1974, Rayonier Canada, the largest forestry company operating on Haida Gwaii, presented a five-year plan to the British Columbia Ministry of Forests to log Moresby, Lyell, and Burnaby Islands. They were targeting the interior of the islands, home to the most productive salmon forests. The Haida, and some islanders and ecologists, disputed this plan and presened a proposal for creating the South Moresby Wilderness Area. In 1979, the summer Jean worked on Moresby Island, the government countered that the economic impacts of *not* exploiting the Haida forests were too great.

The Haida people and environmentalist allies continued to protest. The loss of the forests was starving their culture and threatening irreparable harm to the salmon streams. Rayonier Canada continued to pressure the government to renew its license to log. The government found some "middle ground" by placing a moratorium on logging small Burnaby Island. Rayonier responded by dramatically escalating its clearcutting on Moresby Island. The biggest trees were taken first and shipped to sawmills on the mainland.

In the fall of 1985, the Haida Nation mounted a peaceful rebellion. Haida community members and allies set up a blockade on the logging road in hopes of forcing the Canadian government to recognize Haida law and put a halt to the logging of the last stands of old growth, especially the richest stands around the salmon streams. For three months, the blockaders stood their ground and demanded that their voices be heard. "We will not be pushed aside in our own homeland and told that our interests are not worthy of consideration," said Miles Richardson, then president of the Haida Nation. Seventy-two people were arrested, among them several Haida elders.

The logging resumed, and legal and political controversy continued to rage. But the blockade had had an effect. The public was paying attention to what was happening on Haida Gwaii. After several years of trading proposals and memos back and forth, the government of Canada and the Council of the Haida Nation signed an agreement in 1993 creating Gwaii Haanas National Park Preserve and Haida Heritage Site, safeguarding at least some of the old-growth forests on the southern end of the archipelago, including Lyell, Burnaby, and Kunghit Islands, with co-management. The editorial board of a forestry journal deemed the park's creation a "sad story," enabled by "a tapestry of scientific fact, pseudo-science and mythology."[1] Indeed, only about 15 percent of Haida islands were preserved, the rest still in the crosshairs of colonial interests. But with a portion now protected, Richardson declared that the decision of the Haida Nation had been upheld.

"This is Haida land and there will be no further logging in this area," he vowed."[2]

With the Gwaii Haanas Agreement signed, Graham Island, where most Haida lived, continued to be logged heavily. Timber exploitation didn't stop, it just crept to the north. The logging continued to target the heart of the islands in the archipelago. Even the boggy forests to the north, where the trees were small, and the windswept forests of the rocky west coast, where the trees were twisted by the intense winds, were not safe.

LISA WHITE-KUUYANG, a Haida cedar and raven's tail weaver, had lived on Haida Gwaii her whole life. Her father had been a carver, fisherman, boat builder, and traditional Haida canoe builder. Her grandmother, who had been a matriarch of the Yahgu'laanaas clan, had taught her that it was the people's responsibility to protect the lands and waters for future generations. Born and raised in Old Mas-

Log barge in Massett Inlet, Haida Gwaii, 2021. Timber companies have historically barged almost all the cut logs off Haida Gwaii, mainly at night. Over half of the logs have been shipped directly overseas, and the rest to mainland British Columbia. Little timber had been left on island for the Haida to meet their needs or create value-added processing jobs in their communities.

sett, Lisa had married and raised her family in Daajing Giids one hundred kilometers (sixty miles) to the south, returning to Old Massett sixteen years later.

Her community had watched their forests fall year after year, the trees taken away on gigantic log barges to the mainland or some distant country. Now the primary forests on Moresby Island were all but gone, and those on Graham Island were vulnerable to continued logging. Haida villages were surrounded by clearcuts or second-growth young forest. Unwanted logs were left behind in the battered clearcuts. Lisa knew the logging was causing the salmon, northern goshawks, ermine, and ancient murrelets to disappear. There was so little accessible old growth left that she had to drive for hours to gather the cedar bark she used to make her traditional weavings. She wondered if there would be any forests left for her children and grandchildren.

LISA WAS DESPERATE to find alternatives to clearcut logging that she could share with her community. She had written to invite me,

Lisa White-Kuuyang near Yakoun River on Haida Gwaii, 2021. The western redcedar behind Lisa is thousands of years old and is known by the Haida women as *díi k'wáay*, or Older Sister. The smaller trees are Sitka spruce and western hemlock, and the forest floor is carpeted with dozens of moss species. These forests are home to the endangered northern goshawk Haida subspecies *stads k'un*, meaning "wings brushing bows," which serves as a keystone shaping the entire food web of the forest. The forest is also habitat for ancient and marbled murrelets, northern saw-whet owls, and a large subspecies of Haida black bear known as *taan*. The salmon runs in the Yakoun River have declined precipitously due to habitat loss from clearcut logging, overfishing, and climate change. We found deeper carbon pools in these forests than others reported for coastal temperate rainforests, which appears related to the nutrient subsidies provided by the salmon that had been stewarded by Haida for over fourteen thousand years.

Teresa, and Jean to come to Old Massett in hopes that we could help "stop the chainsaw massacre" of Haida Gwaii's remaining primordial rainforests. This was exactly why we had established the Mother Tree Project—to provide science-based alternatives to clearcutting, help develop harvesting practices that would protect and tend the forests instead of destroying them, and encourage community-based management so local people could collectively make decisions about the forests that sustained their communities and their cultures.

If we could get funding, we could include the forests of Haida Gwaii in the Mother Tree Project, just as we were exploring in Heiltsuk territory, representing the northern limit of the climate gradient along the coast of British Columbia. Like most of the forests along this coast, Haida Gwaii was dominated by western redcedar, hemlock, and spruce, and absent of Douglas fir, which prefers a drier climate. A Mother Tree experiment on Haida Gwaii would therefore focus on protecting the western redcedar mother trees, rather than Douglas fir, as with our existing nine experimental forests to the east. But because the one coastal site we had established at Malcolm Knapp Research Forest protected both cedar and Douglas fir, we could make comparisons with cedar-dominated experiments along the Pacific coast. As our climate warmed, the coastal sites as far north as Haida Gwaii might even become more favorable to Douglas fir.

We knew it would be costly to expand the study. Our research was underwritten by the governments of Canada and British Columbia and a handful of generous grants from private foundations. Cedar was culturally important to First Nations peoples along the coast and played a foundational role in the integrity of these ecosystems. We hoped these facts would help persuade the provincial Ministry of Forests to fund our research.

In my youth, I'd thought of Haida Gwaii as a distant, mysterious jewel. My scant knowledge of this place was shaped mainly by a Haida family, the Mills, whose kids attended my high school. The Mills were just one of the many middle-class families in the northernmost rural Kamloops neighborhood called Westsyde. We kids were oblivious to what had brought this family from Haida Gwaii to the arid woodlands of Kamloops (Tk'emlúps te Secwépemc).

In high school, we'd learned little of the First Peoples or the effects of colonialism. While Mum had taught us from childhood that Indigenous people were the rightful stewards of the land, we knew nothing, in the 1960s and '70s, of the abuses taking place at the Kamloops Indian Residential School, ten kilometers across the Thompson River from Westsyde High School. Our heads were stuffed with English literature and French grammar, yet we were unschooled about the land and its original people.

When Robyn was in grade eight in the early 1970s, she'd played soccer with two Indigenous girls named Lana and Cora, and at their invitation, would go stay with them at the residential school, lining up for dinner with the other kids in the steel-clad cafeteria. Lana and

Cora taught Robyn traditional beading outside in a stairwell. In retrospect, Robyn wondered if they were hiding. She still remembers the sound of children crying at night in their white cast-iron beds in the dormitories. Children who had been forcibly separated from their families and communities.

From my present perspective, way up here in the fog, I now understood far more. The federal government of Canada had long planned for the dissolution and impoverishment of the First Peoples, and this had limited the opportunities for future generations, either forcing or drawing them from their communities. Their territories were stolen and the people oppressed, while settler families like mine remained largely uninformed of the history surrounding them.

Teresa sat in front of me, her eyes darting across the landscape with concern. Her black hair was swept back, and around her neck she wore a Maori pounamu, a jade pendant carved with fern fronds symbolizing rebirth and growth. Her own Ts'msyen people—whose name means "inside the Skeena River, river of mists"—lived just a stone's throw away across Hecate Strait. The Ts'msyen, like the Haida, were also people of the cedar and the salmon. For thousands of years, they had tended to the majestic trees for carving canoes, longhouses, masks, and poles. The bark was used for clothing, baskets, mats and hats, and medicine. The salmon fed the cedars, bears, and people. As a salmon-fisheries scientist, Teresa was at the forefront of scientific investigation into the condition and restoration of these resources. She carried with her inherited obligations handed down through her mother's lineage to care for the land and the resources.

We stared down through the clouds at the two landforms—the eskers and the bogs—that were fitted together in a whole ecosystem. The hydrologic system was synchronized with the rain and the tides, the ocean currents and the mountains. The rainwater gushed from the clouds and wound through the trees, soils, and rivers and back to the ocean, enabling the growth of immense trees. Across the ages, the forest ecosystems had enriched the glacial moraines with salmon and humus, while the sedges and mosses had filled the impervious bottomlands with roots and organic litter. With only wind disturbing the forest, each passing epoch laid down deeper and richer carbon pools. The Haida people were in rhythm with the water flows, thriving over millennia by tending to the salmon and cedars, connections and cycles. This was one of their teachings: that human societies lived in symbiosis with the land.

Teresa turned to me in dismay. The land had come into sharp focus as we emerged from the river of clouds. What had looked deeply green at 3,500 meters (11,500 feet) was now visibly pocked with geometrically shaped clearcuts. The cedars and hemlocks had been stripped from many of the eskers, the waste wood piled in apartment-sized towers. The old-growth forests I had imagined cloaking these islands had either been logged or were in imminent risk of being clearcut.

As we started our descent into the township, the wind buffeted the tiny plane like a kite. The plane plunged into a bank of fog, the air banging its wings. Then it hit a high-pressure bubble like a speed bump, followed by another plunge. I white-knuckled the seat arms. Sparsely treed bogs popped from the fog. The air lifted the plane a little more gently. We were enveloped in another fog bank. Then the wheels bounced along the gravelly pavement of a short runway until we slowed and taxied to a small wooden building.

LISA MET US at the Massett Airport and took us straight to one of the remaining uncut forests a half-hour drive to the south. A dozen people—Haida, Ts'msyen, Ktunaxa, settlers, loggers, teachers, and foresters—were waiting at a small trailhead, all comfortable in their rain gear, joking about the weather. One Haida man chuckled that he was going to lose his rubber boots in the bog, and everyone laughed at my flimsy runners. But beneath the jokes ran a deep vein of worry.

We walked along a narrow path through sphagnum hummocks toward a wooded esker. The open bog canopy offered little protection from the dripping clouds. Lisa pointed out the tiny crimson bog cranberries, the orange-red cloudberries, and woolly Labrador tea. A sandhill crane stood Jurassic in the mist.

The trees in the bogs were small and distinct from the cathedral forests among the fjords and islets of the mainland to the west and south. The impervious soils, compacted by glaciers eons ago, now supported bog meadows intermixed with forest. Oval-leaved blueberry and crowberry nestled among the sphagnum mosses, and the seemingly bottomless organic soil held great pools of carbon accumulated over thousands of years. Instead of the productivity aboveground in enormous majestic trees, we figured it must lie underground in massive soil carbon storage banks. This worried me and Jean. The deep bogs, entwined with the moraines, would be at risk of degradation with the logging. Once the forests on the eskers were cut, nutrient-rich waters would flow down into the fens, possibly disrupting the hydrology, enhancing decomposition, and increasing respiration.

Under a canopy of mostly slender cedars, the members of the group voiced their concerns. The government had set excessive quotas of timber to be cut. Market-driven policies were behind extensive clearcutting, with raw logs shipped on barges overseas. Timber companies made large profits, valuing the wood for the same durability and lightness as the Haida did. Little profit flowed back to the people on Haida Gwaii. Most of the monumental cedars had been replaced with yellowish Sitka spruce, lodgepole pine, and western redcedar seedlings, and the cedar regeneration was heavily browsed by the invasive Sitka black-tailed deer. These non-native deer had been introduced in 1878 to supplement the settlers' diets. The deer had completely transformed the forests by browsing the understory plants and regenerating trees throughout Haida Gwaii. The remaining seedlings were also stressed from the unusually warm summers. The people feared that cedar would soon disappear from the big islands.

The mood had turned somber.

"Haidas had respect for every living thing. They never took more than they needed or more than nature could provide," Lisa said.

A quiet woman who looked about my age spoke up. She said she was a primary school teacher who held a master's degree in education.

"The ancient cedars are our grandmother trees. We honor them because of their longevity and wisdom, but the government agencies ignore us because they think we're not scientific enough." She asked if Jean and Teresa and I could speak to the government bureaucrats in a way they would understand.

I summoned my thoughts.

"These old cedars are similar to the mother trees I have studied in the Douglas fir forests of interior British Columbia," I said, pointing to a grove of century-old cedar trees whose lower branches arched down to kiss the ground. "The older trees are connected to the young saplings through an underground system of roots and fungi. But unlike the Douglas fir forests, where seedlings only regenerate from seed, the branches of cedars become new saplings in a process called layering."

When I pulled at a low-slung branch diving down from the fluted bole of a parent cedar tree, it didn't budge. The underside had sprouted roots, anchoring it to the soil. The swooping branch looked like an elbow leaning into the earth. Fallen leaves and decaying mosses had gradually layered over the rooted branch until the crook was buried. I dug into the feathermosses, carefully removing the layers of decaying leaves and stems. I picked the *Hylocomium splendens* rhizoids from under the woody elbow and exposed a row of sturdy red roots that had established firmly in the soil. A dozen crimson arrows shot straight down into the minerals. From the roots, the branch had swooped upward and away, then straightened toward the sky like a new sentinel. What had started as a lower lateral branch was turning into a separate tree in its own right. The buds of what once were lateral cedar plaits would soon turn into the primordia of a new tree. In time, having grown its own crown providing sufficient photosynthetic energy for maintenance and growth, the daughter would individuate from the mother and become an independent sapling.

Additional saplings had encircled the mother tree, indicating other branches had already separated as individuals. The teacher nodded, smiling. The mother tree had supported her daughters through her connections until they could feed themselves. In time, as the tree aged, she would spawn multiple generations of saplings in tune with light gaps emerging and weather turning favorable, and this history would be stored in her tree rings. As she grew older, she would become what the teacher called a grandmother tree.

Looking through the patch of forest, we could easily spot the daughters and granddaughters growing in halos around their elders. Shaggy-barked yews were sprinkled among the cedar circles.

In the worldview of the Haida and the Ts'msyen, the cedar tree was the older sister who connected the land and the sea. A generation of cedars in a patch were called sister cedars.

In her doctoral research, Eva would sequence the DNA of the arbuscular mycorrhizal fungi of cedar, yew, and maple in the inland rainforests and find that these three tree species shared an abundance of arbuscular fungal species in common. There were no maple trees on Haida Gwaii, but there were plenty of cedars and yews. The forty mycorrhizal fungal species that overlapped these trees likely connected them together with their fine fungal threads. Eva would find that the trees were a part of a tight-knit community. The overstory—whether cedar was dominant or mixed with hemlock—helped shape the diversity of fungi in the shared mycorrhizal network. Yew, the wizened old tree of the understory, was especially influenced by the identity of its overstory neighbors. This made sense, given that the meandering yew needed overstory trees to thrive and was tightly attuned to the spatial distribution of its taller neighbors.

"Do the cedars and yews connect with huckleberries?" the teacher asked.

These three species were essential to Haida people for clothing, tools, medicine, and food. I described how the huckleberries formed ericoid mycorrhizas that involved different fungal species, but that some researchers had found arbuscular mycorrhizal fungi in huckleberries elsewhere in the world. The soil microbiota of cedar, yew, and huckleberries growing together would certainly be shaped by the communion of these companion plants.

I could see how profoundly connected the Haida people were to their land, and how shattering the loss of the great cedars would be.

We traveled to a nearby clearcut. I gazed at the wind-tossed timberline and felt sick.

"They clearcut the cedar, then meet minimum government regulations by planting spruce, pine, and cedar," a logger said with a sigh. "It looks like a graveyard."

The cedar seedlings were planted inside white plastic tubes, intended to protect the young trees until they were tall enough to survive foraging by the Sitka deer. But cedar couldn't reproduce well using these industrial forest regeneration practices. Not only did cedar regenerate mainly through layering from the mother trees, the saplings needed overstory cover for protection against sunscald. Even the lady ferns that had managed to live through the cutting operation were yellowing from sunburn. With the certain death of the arbuscular mycorrhizal fungal threads in the absence of the cedars, yews, and ferns, this site was at risk of not regenerating back to cedar. And without saplings growing into parent trees, the next generation of cedars would also be in jeopardy.

The method of regeneration was only a fraction of the problem. Our group walked through the clearcut along an oily road underlain by trucked-in blasted rock. The surface was covered with glacial till dug from pits within the cutblock. These roads amounted to 7 percent of the forested area, as regulated by provincial law, and would never have the capacity to grow old trees again. The swimming-pool-sized pits had been excavated every one hundred or so meters along the road, draining the rich soil of tannin-darkened water and speeding up decomposition of the forest floor. This practice would have made planting the pines easier.

But the main purpose of the pits was to serve as dumping grounds for thousand-year-old yellow cedar logs. Cut alongside the redcedars, these yellow cedars were discarded because they were not profitable in the market or were too difficult to mill. The spiral grain of the yellow cedars, a natural adaptation that strengthened the tree against the

Seedlings planted among slash following clearcutting, Haida Gwaii, 2021. White plastic tubes are placed over seedlings, grown in nurseries off island, to protect them from browsing by the introduced Sitka deer.

strong Pacific winds and helped deliver sugars to deeply inquisitive roots and branches, messed up the head saws. Adding more insult, slash piles nearly thirty meters tall were full of old-growth redcedars that had been considered waste because of some "defect"—a twist, a fork, a rotten center, and the tree would be tossed into the pile. Lisa told us of seeing huge burn piles in remote parts of the islands, stuffed with the remains of old-growth trees. Loggers were burning and burying trees that had taken centuries to grow.

"It's barbaric. A blatant waste of our precious trees," she said.

The greenhouse gas released by the clearcutting of the old trees was changing Haida Gwaii from a globally crucial carbon sink into a carbon bomb.

The Haida people gathered in the high school gymnasium that evening. Teresa spoke about her hereditary connections to the Haida, and the science of the Mother Tree Project. The project resonates with Indigenous people, she said, because it emulates the connections observed in traditional knowledge. She quoted the words of Kwakwaka'wakw chief Robert Joseph, who said, "When everything is connected, we remember balance and harmony."

I gave a presentation on our research on mother trees connecting seedlings and plants through underground webs of mycorrhizal fungi, and drew comparisons between the social lives of trees and the organization of human societies.

Jean spoke last. I knew how difficult it was for her to stand up in public. She preferred to work in the background, keeping good relations with her peers and students, always being thorough and careful and only speaking when it was time.

Now was the time.

She moved to the front of the audience, her greatest fear. Her gait was steady, revealing no signs of her MS, but her voice trembled with emotion. I wanted to run up and stand beside her, to support her like a sister cedar. But she went on, her words measured and thoughtful. She recounted her experience working for a forest company on Haida Gwaii when she was twenty, and how working among the massive cedars had changed her life. How the energy, clarity, and spirit of these forests at the meeting of the land and sea was unlike anything she'd ever seen. She shared the awe she'd felt walking through the Windy

Bay watershed that summer. Her eyes welled up as she described the industrial clearcutting of the Haida forests. How she'd once helicoptered to a small island that had been completely shorn of its trees and seen a lone bear left bewildered on a landing.

Jean admitted that, at the time, she hadn't understood the impact of the logging on the Haida Nation. She said she was honored to be here now working in support of the Haida people, helping to bring scientific methodologies to support better ways to manage the forests. When she finished, the room was silent, thrumming with purpose. We all knew we had a role to play in protecting the remaining old-growth forests of Haida Gwaii.

That summer, Teresa and I returned to Old Massett to join the Haida in raising a totem pole, or *gyaa'ang*, at Tlielang on the northern spit of Graham Island. With so few monumental cedars remaining, the six-hundred-year-old tree had taken two years to find. Over six months, master carver Kihlguulans Kihl 'Yahda Christian White, Lisa's brother, had carved the fifteen-meter trunk into a welcoming pole with the help of eight young apprentices. The project to carve and raise the *gyaa'ang* had employed eighty people all told. Teresa and I cheered with the islanders as the monumental pole was raised in a sunny clearing beside the Hiellen River—the first pole raising there in two hundred years. The pole told the story of the people and the land, with carvings of ravens, eagles, beavers, and bears gazing back at us. This *gyaa'ang* would stand for decades, joining the other cedar poles that still stood sentinel around the islands.

Lisa told us that the poles used to be cut down and burned by order of the church ministers, who told the Haida people that they wouldn't go to heaven if they kept their totem poles up. The Christian ministers misunderstood the significance of the totems. They thought the Haida worshipped or prayed to them. In fact, the *gyaa'ang* were living records of a family's history and depicted their crests and lineage. Other poles were stolen from the deserted villages on Haida Gwaii after the smallpox epidemic killed over 95 percent of the population in 1862. For years, the carved poles were also removed by anthropologists and collectors, taken or sold for display in museums and galleries in the belief they were being saved from deterioration. Yet that was another fundamental misunderstanding of Haida culture. The poles

Lisa White-Kuuyang observing remains of western redcedar, Tree of Life, in a clearcut on Graham Island, Haida Gwaii, 2021. About a quarter of a logged forest ends up in forest products in British Columbia. Around half is reduced to sawdust at the mill, and the rest is left as slash on the site. The forest industry does not consider the opportunity cost of this practice to future generations.

were intended to decay back into the forest from where they came, following the natural cycles of all living things.

I knew that the ancient cedars were much more resilient than a few decades of poor forest management. In time, the sister cedars would prevail. The Haida people were equally resilient. They were teaching us to heed the laws of nature.

"This is our homeland," Lisa said. "They can take all of our trees, but we aren't going anywhere."

A COUPLE OF MONTHS LATER, Jean, Teresa, and I wrote a proposal to the provincial government requesting funding so we could document the effects of clearcutting on the carbon pools and plant diversity on Haida Gwaii, and test selective methods of harvesting some trees while maintaining the integrity of the forest. In the spring of 2018, our proposal was rejected. The head of the program wrote that there was already sufficient partial retention research in that area. Besides, clearcutting was the only harvesting practice used in coastal forests. The provincial government was not open to considering other practices that would leave valuable old trees standing. Our hopes of establishing a Mother Tree Project experimental forest with the Haida people had evaporated. We were stung with disappointment, but logging, planting, and sampling at our established nine experimental sites kept us occupied for the rest of the year.

In the late fall of 2019, Lisa emailed to tell us that there was an opportunity to comment on the timber supply review for Haida Gwaii. This was a crucial chance to convince the chief forester of British Columbia to reduce the rate of clearcutting on Haida Gwaii. Less than 3 percent of British Columbia's most productive old-growth forests in the working landbase remained, and this was probably the last chance to stop the felling of the remaining ancient cedars. Lisa hoped our input could persuade the council to consider alternatives besides clearcutting, or better yet, support a moratorium until an independent assessment of the health of the lands and rivers could be done.

Teresa and I wrote our letter on New Year's Day, 2020. We marshaled our facts on the loss of carbon stocks with clearcutting we'd documented elsewhere in the province, and the dangers of runaway climate change, the damage to biodiversity and endangered species, the harms to forest hydrology from careless logging practices, and the detrimental impacts to wildlife habitat. We recommended a moratorium be placed on any further clearcutting on Haida Gwaii.

The same day, Lisa submitted her own letter regarding the timber supply review. Into this letter she poured generations of knowledge about her ancestral home and its laws. She questioned why the cedars were being replaced with spruce and pine seedlings from mainland nurseries. "This is not acceptable," she wrote. "Why is the growing of Haida Gwaii seedlings not being done locally?" She described how the timber supply review failed to address harm to the countless species of plants and animals "that have equal importance, not only to the Haida, but the biodiversity left in the world." Poaching of both trees and bears was going unpunished. Nor did the review address record low salmon numbers. During a weekend opening of the Haida food fishery, it appeared that only four fish had returned to the Yakoun, the biggest salmon-bearing river on the islands, shocking the islanders.

A buffer of one and a half tree lengths was now required around

Clearcuts on Haida Gwaii, 2021. The most productive salmon forests have been logged first, leaving less productive forests in fringes, on mountaintops, among bogs, or on cliffs. The companies are now required to leave a buffer of one and a half tree lengths along streams and rivers to conserve salmon habitat, but as climate changes, this may not be enough to protect the fish. The fringe of trees along the shoreline serves to mask the clearcuts from the view of tourists boating by the islands.

streams to keep the heavy logging machinery from degrading the stream banks. But with increasing storms due to climate change, Lisa pointed out that those buffers might not be sufficient for long, resulting in the river being flooded with silt, degrading the habitat the salmon fry needed to survive. She wrote that visitors to the northern part of the islands were often shocked by the extreme level of cutting and waste visible just off the highway, and she worried this would affect the community's ability to attract more visitors and create jobs in the tourism sector.

"Trees are gone once they are cut," she concluded.

Our letters made their way to the provincial chief forester through the Haida Gwaii Management Council. Now there was nothing left to do but wait. I knew it was unreasonable to expect a quick response from government, but with thousand-year-old trees being felled every day on Haida Gwaii, the matter was urgent.

Back home, I looked out my window at the Selkirk Mountains.

Snow was falling on the rampant forest, the cedars, hemlocks, and Douglas firs cloaked in thick crystal jackets. These forests gave my little town of Nelson security, spirit, and joy. But brand-new clearcuts blazed starkly against the slopes and ravines, bright and rigid. The absence of forest cover in these openings would lead to more snow on the ground and more solar radiation reaching the snowpack. With the arrival of spring, the snow would melt rapidly, and the forest floor, missing its absorbent carbon sources, and the tree roots, now dead, would be unable to sop up the meltwater. The water would rush down the steep creeks and possibly destabilize the alluvial fans, like the one on which my house was built. Clearcutting had already been identified as the main cause of landslides and flooding in the mountainside towns of British Columbia, and Nelson could be next on the list.

In May 2020, five months after we'd submitted our letters, the chief forester wrote that the timber harvest rate on Haida Gwaii would remain 90 percent of the previous cut. This harvest level, they said, was sustainable. The clearcutting would continue unabated.

8

CARBON BOMB

The Southern Interior Silviculture Committee, or SISCO, is a big deal for forestry in British Columbia. Though focused on the southern interior, the conference is attended by people from all over the province and from all walks of the forest sector, from registered professional foresters, forest technicians, and university academics to government policymakers, forestry consultants, and employees of private logging companies. I'd first presented at SISCO when I was twenty-nine, right after finishing my master's at Oregon State University, and I still attended whenever possible.

This year I would be giving a presentation on the Mother Tree Project at SISCO's winter workshop, the first week of February in Kelowna, a city in the southern interior on the shores of Okanagan Lake. Jean and I had been working feverishly to compile my presentation, which had the cumbersome title "Carbon Storage and Biodiversity Across a Climate Gradient in Interior Douglas-Fir Forests, Before and After Logging (Clearcutting and Partial Cutting)."

I drummed my fingers on the windowsill and watched a hawk circle over a squirrel in the branches of the maple outside my window while I waited for Jean to crunch the data and send me the figures. Finally, my computer made its familiar *ping*. I opened the new data Jean had sent and stared at the graphs on my screen, stunned. At each of our experimental forests, across our climate rainbow, from Twaal Creek and Jaffray to John Prince and Malcolm Knapp, the carbon stocks aboveground—the community of trees, shrubs, herbs, mosses, lichens, liverworts—had disappeared with clearcut logging. The species and their habitats had been obliterated by the machines. But it wasn't that making me speechless.

My hands were trembling as I dialed Jean on Zoom, and I waited impatiently while she fiddled with her video and audio settings. Finally, she appeared on the screen.

"Jean! Did you notice the forest floor?"

She'd been so busy making the graphs that she hadn't had time to think about what they signified. We checked the numbers and reran the analysis, just to make sure. While the programs were running, Jean grabbed another cup of coffee. When the analysis was complete, we stared at each other in disbelief. The logging had caused a *61 percent reduction* in forest floor carbon. This was over *five times* the losses detected in other analyses of temperate forests around the world. *Holy shit!* Could we have made an error in the calculations? I reread the figures and checked the data columns again, but the figures came out the same. While I'd expected to discover some degree of loss, this was far worse than I'd anticipated.

"You're right, HH," she said, "most of the carbon in the forest floor is gone."

Over half of the ancient underground carbon pool that had taken thousands of years to accumulate had instantly disappeared with the logging. Even more stunning, the losses from the forest floor were just as acute in plots where some trees had been left behind as in plots that were completely clearcut. Not just the organic matter, but also the soil organisms living in the forest floor would have also lost their habitat. Most of the fungal species making up the mycorrhizal networks would have perished along with the trees, and any remaining threads left behind diminished, fragmented, and decaying.

"You would think that leaving patches of old trees would have protected some of the forest floor," Jean said, as we puzzled over the numbers.

We double-checked the data from all of the treatments one more time, but the data analysis was correct.

"I hate to say it, but I think the machines are the problem," I said.

The logging equipment was a major factor distinguishing all of the cutting treatments from the uncut control. Machines like the enormous feller bunchers, skidders, and forwarders weighed up to 25,000 kilograms (55,000 pounds) each and could crush mature trees. As they crawled through the forest, their massive tracks or wheels would grind, displace, and compact the forest floor, in some places squeezing the water out of it like a sponge. Even in treatments where different levels of overstory were left behind, the feller bunchers would still wind among the trees, leaving treads throughout the stand.

"It's possible we're not seeing a difference between the clearcut and partial cuts because we need more sample plots," Jean said.

There should have been some protection from the residual trees, but the shoving, piling, and scraping would have created even more mounds and dips in the forest floor than naturally occurred. This could have resulted in much higher and lower forest floor depths than normal, making it difficult to find statistically significant differences between the nuanced logging treatments. Sampling more microplots, where we quantified carbon in the soil pools, in each retention patch would increase the precision of our forest floor estimates.

"I have to say, though, that the forest floor in the clearcuts *looked* so much worse than in the high retention treatments," I told Jean.

When I'd walked the barren clearcuts with Eva a few months back, with Tia bounding at our heels, I had been stunned at the extent of the scrapes and gouges in the earth caused by the logging operations completed the previous fall.

Jean had been having trouble walking on uneven ground, so she'd left it to the rest of the Mother Tree Project crew to complete the post-harvest assessments. She missed the fieldwork, but she found ways to get out whenever she could—working at roadside plots, using a walking stick, taking notes from a stump. Her illness had not dampened her spirit for the forest in the least.

Now Jean and I reviewed the photo library—each picture methodically filed in the UBC computer system. The images emphatically reinforced what our data suggested: the forest floor had been severely damaged by clearcutting, leaving extensive areas of exposed mineral soil. Some microsites had been severely impacted, while other patches around protected trees were still intact. But the forest floor had also been diminished in the partial retention treatments.

Even with these dismal results, I saw a few reasons to be

encouraged—if I squinted. The fraction of aboveground carbon remaining in the partial retention treatments matched the proportion of overstory trees spared by the saws. Where 10, 30, or 60 percent of the trees were saved, 10, 30, and 60 percent of the carbon remained standing in the aboveground pools. Many of the understory creatures also survived, sheltered among these protected trees. Wherever there were still branches to nest in, trees to den in, bark to grow on, and forest floor to root through, the birds, bears, lichens, mosses, shrubs, and herbs had endured.

We decided to double the number of forest floor samples in the coming summer months to help us see if there were significant differences between the clearcutting and the partial retention treatments. This meant our crew of eight students would return to all nine forests and document all of the carbon pools again—this time in eight microplots per harvesting treatment instead of just four. It would take our crew two months to take these additional measurements. Then we'd know if the similarity in forest floor carbon stocks among the clearcut and partial retention treatments was really due to similar intensities of machine traffic, or if we were missing subtle differences due to insufficient sampling intensity.

But I knew this did not really matter much in the grand scheme of things.

Clearcutting was undisputedly the easiest and cheapest method of harvesting trees. It required little knowledge of the forest or understanding of the ecosystem. I knew that it took more effort and ingenuity to leave some trees behind than to take everything.

The last time my province had reported silviculture systems statistics, in 2018, over 94 percent of the harvesting had been done with clearcutting or else what was called "clearcutting with reserves." Clearcuts, by definition, are areas where at least 50 percent of the logged area is no longer influenced by a forest edge and have a minimum size of two hectares. Clearcutting with reserves differed only in that small patches of trees were left inside the clearcuts. Clearcutting with reserves, in my view, was really a kind of greenwashing, meant to appease an unobservant Canadian public and distant international audiences that British Columbia forestry was more innovative than it really was.

Mother Tree Project crew hauling soil samples from an old-growth cedar-hemlock forest in Sinixt, Ktunaxa, and Syilx territories in the Columbia Mountains of interior British Columbia, 2023. They would find that carbon stocks in these wet belt forests were the highest measured in the interior of the province. The thick-barked tree in the foreground is western hemlock, and the tree with stringy bark to its left is western redcedar.

But anyone driving the highways or flying over the province could see that the tiny reserve patches inside clearcuts usually contained lower value trees, shocked by the loss of their neighbors. Located on rocky outcrops, along steep slopes or in swamps, these remnants functioned poorly as forest reserves. The trees did not provide quality seed, protect the understory, or provide much habitat for wildlife.

I thought of Lisa's letter to the chief forester. She did not mince words about the practice of clearcutting with reserves on Haida Gwaii:

> How do bears live in these areas, or do they have to be told to go live on the forest "reserve"? Knowing that bears have poor eyesight and are dependent upon their ancient pathways to navigate their territories and need old growth to live in, makes me think that they too will become extinct over time. "Reserves" don't work for people and they certainly will not work for animals as it's proven they only isolate and impoverish.

The loss of carbon from the forest floor with clearcutting was shocking when added up at global scale. Logging, along with other forms of land-use change the world over, has been estimated to account for over 20 percent of greenhouse gas emissions. These estimates were probably underestimates, I thought, since they didn't account for this new data.

The loss was also *irreplaceable*. It had taken several millennia of carbon cycling—the food web of creatures working together to decompose, transform, and concentrate the organic compounds—to store this ancient detritus under the ground. In the week it took the machines to scrape a stand of trees off the surface of the Earth, thousands of years of sequestered carbon was pushed with logs into piles and burned, combusted, vaporized, or eroded away. At the rate climate was changing, and in the time the world had to increase carbon sinks—instead of releasing more greenhouse gases to the atmosphere—the forest would not be able to recover these stocks.

Jean and I stared at each other numbly. We were holding evidence that Canada, through its forest practices, was detonating a carbon bomb. When it came to carbon pools, *logging was degrading the forest*. Land degradation, like deforestation, was a serious condition that threatened our future on this planet. A degraded forest could no longer filter air and water, sequester carbon, or provide animals and people with food and shelter as effectively as it once did. Some of the trees might still be there, it might even continue to resemble a forest, but it was a shell of its former self, and it could take centuries to recover.

Clearcut with reserves on Haida Gwaii, 2021. In this method, the forest is clearcut and a few small patches are retained. These patches are said to provide habitat for wildlife, disperse seed for regeneration, protect monumental cedars from blowing over, and mitigate the visual impact of clearcutting. Often, however, the retention patches are too small or poorly located to meet these goals.

Only a few days earlier, at the COP26 conference in Glasgow, Canada had joined 136 other nations in a pledge to halt and reverse biodiversity loss, deforestation, and land degradation by 2030. The Mother Tree Project data suggested our country was already breaking its promise with respect to land degradation. Jean and I would have to publish the data quickly.

THREE DAYS LATER, I packed my overnight bag, dashed out the door with my jacket flying, and drove over the snowy mountain passes to Kelowna. By the time I arrived at the hotel where SISCO was being held, the preconference soiree was already well underway. People were cheerfully greeting each other, happy to see colleagues after a long winter. I had a beer with some old friends and we exchanged stories about our children and our latest adventures, and then I slept restlessly, the data swirling in my head. I wished Jean could have been there with me, but she'd had too much to do at home.

The next morning, people slowly filed into the room, and by the time I was ready to start, every seat was filled. I began by describing the Mother Tree Project, our partners, and our research methodology. I slowly clicked through the slides, giving the presentation Jean and I had written. When I showed the graph comparing the effects of our different retention treatments on the carbon pools, the room went silent.

"Where did the forest floor go?" a company forester asked.

I could see people leaning over to whisper to one another. The realization that clearcutting—our standard practice long considered sustainable, and certified as such—was causing such damage to the forest was just starting to sink in.

"Do you think the soil organic matter was just moved around the site by the machines?" a young consultant in wire-rimmed glasses asked. Maybe it was piled in some corner of the woods, like coins in a bank account. But these would be the huge slash piles left on landings and along spur roads. They were normally burned in the fall, the humus combusting along with the unwanted logs and branches when lit with a drip torch. Sometimes they were just left to rot.

"Maybe the forest floor has been ground into the mineral soil," suggested a senior forester, a pile of notes stacked on her lap. But our sampling methods would have detected fragments of forest floor in the surface mineral soil layer—the A horizon in the soil profile—and they simply were not there.

A government forest scientist in a red knit cap said, "I bet it leached into the subsurface horizon—the B horizon—with the fall rains."

But this explanation wasn't supported by the data either. We found no additional carbon in the B layer. Besides, translocation of carbon into deeper layers was a long-term process that takes place during the formation of soil, not something that occurs over a few months.

"Is it possible you accidentally included mineral soil in your forest floor samples?" a soil scientist asked me. This could point to inflated carbon stocks in the control forests, he ventured, setting up the possibility for accentuated contrasts with the logged treatments. I explained that we had followed standardized methodologies to prevent this, knowing that accidentally including some heavy mineral particles in the lighter forest floor samples could make the organic layer appear more carbon dense. In the soil profiles of interior forests, however, the organic and mineral soils are as distinct as the tiers of a layer cake, making it easy to sample with little error.

Besides, we used the same exact sampling methodology in all of our treatments, eliminating the possibility of bias toward controls, or any treatment for that matter. If anything, this kind of mixing would be greater where the machines worked, increasing the possibility of overestimating carbon stocks in the logged treatments, not the controls. And at the end of the day, the carbon stocks we were finding in our unlogged control plots aligned with other published data.

It was crucial to question and review the methodologies, as science depended on peer review and verification. More importantly, lives and livelihoods could be at stake. If our data was correct, we needed to stop clearcut logging in order to prevent further forest degradation and the acceleration of climate change. Such a step could result in closed mills, lost jobs, shuttered towns. Communities would need to develop other methods of managing the forest if people were to continue making a living in the forest. I thought of Grampa Henry with his horses on Simard Mountain, hand-falling single trees, hauling the logs to the flumes, and sending the booms down the chucks to the little sawmills along the river. We couldn't turn back time or bring back horse-logging in any meaningful way, but surely we could be more careful and reduce consumption of our forests.

We need to reduce mechanization in forestry. Of course, the forest industry had taken the exact opposite tack in recent decades, steadily moving to increase mechanization of logging, just as farming had progressed from shovels to combines. Each machine had replaced six forestry jobs to reduce labor costs and increase competitiveness in the marketplace. I could just imagine industrial foresters guffawing at my suggestions. *Wouldn't that go over well,* I thought ruefully.

Partial retention logging, where certain trees are carefully selected to leave behind while others are cut, depending on the goals, would require more logging personnel, not fewer, and this would create livelihoods for people working in the forest. The industry, focused on efficiency for its bottom line, always argued that selective harvesting cost too much in added time and skills of the forest technicians, and resulted in lost timber revenue. But as Chief Dave Walkem had noted, even more complex harvesting systems aren't rocket science, and the costs should not be prohibitive. The value gained in healthy ecosystems, vibrant local economies, an engaged public, and respect for the values of the First Peoples as well as their allies would far outweigh any lost profits for distant shareholders.

I CLICKED THROUGH the rest of my slides showing the treatments and the six thousand data points Jean and I and our students had collected to evaluate the forest floor. Everyone brightened when I showed the shady mature forest controls, and nodded with familiarity at the clearcuts. I lingered over the partial retention treatments because their complexity was unfamiliar, allowing the audience time to make their own evaluations. Our data was solid and deep and we'd consulted with university statisticians during the analysis process. I was impressed with how receptive the foresters in the audience were. They cared about forests and were concerned about the climate crisis.

Foresters had dutifully reforested clearcuts over their careers, thinking the ecosystems would recover fully over a century-long rotation. Here I was telling them their companies' practices were causing degradation of the soil, and hence the forest. *How had this been missed?*

This was hard even for me to grasp. If logging was degrading the forest floor, and clearcutting was removing the precious carbon pools that kept our global biogeochemical cycles in balance, then standard forestry practices were helping destabilize the climate.

I explained that most of the losses would have occurred during the logging event—we'd measured the impacts within a year of logging the plots. The carbon loss occurred because the forest floor, the foundation of the forest, had been scraped, pushed, mixed, and displaced by the logging equipment. Some of the soil organic matter ended up in the slash piles, which were later burned or left to rot. The soil surface would have warmed, now that the forest was gone, speeding up decomposition. The rapid regrowth of nitrogen-rich herb and grass litter would also have primed the soil food web—like adding manure to your garden—enhancing decomposition of the soil organic matter even further. But it still puzzled me why studies done twenty years earlier hadn't detected this degree of loss.

Then it struck me: logging practices and the machinery they employed had changed in the years since scientists last examined the effects of clearcutting on forest floor carbon. This realization came at a thesis defense, where I'd been an external examiner for a doctoral student who'd compared the economic efficiency of different harvesting machines used to clearcut steep coastal forests. These were among the last uncut old-growth stands left on the Pacific coast, as most of the highly productive valley bottoms had been razed over at least once. These remnant steep forests were cut by a machine called a hoe chucker—a tracked machine with a winch, cable yarder, and grapple. The hoe chucker would grasp the felled timber with the grapple, then "chuck" the cut tree down the steep slopes onto landings. A single hoe chucker weighed 37,000 kilograms (81,500 pounds), the equivalent of an adult gray whale.

The student had mounted lasers on different parts of the hoe chucker to trace the procedures moving the wood from the stump onto the logging trucks in order to identify which movements made the operation as quick and cheap as possible. A video depicted the whole process, the men operating their huge machines to take down the forest. The video showed entire trees—boles, branches, and needles—being chucked onto the landings and roadsides. Whereas previous practices had cut branches off where the tree had been felled so that only tree boles were removed from the forest, *whole trees* were being removed and delimbed at the landings. The sound of the grinding machines and breaking trees was unbearable. I felt queasy.

Scientists had issued warnings against whole tree logging over three decades earlier when studies found that this practice caused unsustainable declines in soil nutrients.[1] I remembered Hamish Kimmins teaching this in 1981, when I was completing my undergraduate degree in forestry at UBC. An award-winning professor and a giant in the world of Canadian forestry, Dr. Kimmins had lectured widely on the ecological costs of whole tree logging. But without eyes on the ground, the practice had crept back into the logging industry and become the norm. In place of hand fallers delimbing the timbers in the forest, leaving the nutrient-rich foliage and branches on the ground, huge skidders drag the entire cut tree, stem and crown intact, to the landing for processing. This not only saves the company money in operating costs, it has the potential to provide small biofuels for the rapidly expanding bioenergy industry.

Whole tree removal has gained favor not just because it is cheap and easy. It also reduces fire risk because logging operations otherwise leave a lot of slash and debris on the ground. With the changing climate, practices that minimize fire hazard after logging are now being promoted. But the fifty-ton machines dragging huge tree crowns through the forest further displace and destroy the forest floor. The 11 percent carbon loss from forest floor with bole-only harvesting found in previous studies has ballooned, according to our research, to 61 percent loss with whole tree removal using modern-day industrial machinery.

The foresters at the silviculture conference kept asking questions. *What about the mineral soil?* I explained that so far we had found no loss of carbon from the underlying mineral soil. Mineral soils were generally protected by the forest floor, so they were less subject to the direct physical effects of harvesting. The water content in the soil would also have kept the soils cool and less prone to decomposition.

Rubber-tired grapple skidder (left) and tracked feller buncher (right) at Giveout Creek in Sinixt, Ktunaxa, and Syilx territories near Nelson, British Columbia, 2021. The feller buncher grasps the tree and fells it with a hydraulic saw, then uses the specialized arm to bunch the logs together. The grapple skidder grabs the bunches of logs, lifts them off the ground, and drags them to a central landing. The skidders are articulated in the middle so they can easily maneuver around standing trees. The rubber tires are easier on the soil than the metal tracks, but are limited to 30 percent versus 50 percent grades.

Five years down the road, master's student Oliver Heath, my nephew, would discover that organo-mineral compounds in the B horizon may be showing signs of destabilizing three years after clearcutting, quite the opposite of the suggestion from the fellow in the red cap that more carbon would accumulate in the deeper layer following removal of the trees. With his professors J. T. Cornelis and Les Lavkulich, Oliver sampled down to one meter depth at the Mother Tree Project's Malcolm Knapp, Kootenay Lake, and Jaffray sites. In a preliminary investigation, the pool of organic carbon in the B horizon of these sites seemed to be increasingly made up of fresh, decomposable organic matter, rather than organo-mineral compounds.

The forester with the laptop asked how widespread the loss of forest floor carbon stocks was.

"The proportion of loss is the same in all nine of our experimental forests," I said. It didn't matter if we were in the arid forests of the interior or the humid forests of the coast, we were still removing all of the aboveground life and three-fifths of the carbon from the forest floor with clearcut logging. The machines were just as destructive to the organic matter no matter the forest type.

"But the *total* losses are far greater in the wetter forests." I flipped to the next slide showing three times greater total carbon loss from our most humid woods compared to the most arid forests along the Mother Tree Project climate gradient.

"The more productive the forest, the greater the total loss."

"How long will it take for the forests to recover?" a young Indigenous forestry tech asked. I stared at my feet. I wished I could run into the forest, fill my lungs with the sharp, misty air to escape the frightening reality that was emerging from our data. To help answer the man's inevitable question, one of my students, Alyssa, had used an open-source modeling tool called the Carbon Budget Model of the Canadian Forest Sector to project the recovery of carbon stocks from the different harvesting treatments.

None of the logging treatments had recovered their carbon stocks in fifty years.

The clearcuts were projected to contain 36 percent less carbon half a century in the future than areas where forests were left untouched. This was disastrous. Our data was predicting it would take centuries for primary forests to recover their carbon pools after clearcutting. We were counting on forests to *increase* their carbon sink strength in order to meet greenhouse gas targets, and here we were showing that the managed forests were *losing* their capacity to sequester and store carbon.

Halting the practice of clearcutting and using less destructive machinery was the best option—next to leaving the primary forests uncut—for protecting carbon stocks and thus mitigating future greenhouse gas emissions. Unless of course the forests burned down or became infected with diseases, and this would require careful tending to mitigate such losses. Even so, forests experiencing such natural disturbances are usually left with ample living trees to protect and regenerate the ecosystem.

"Let's fix this problem as soon as possible," one of the foresters said.

The room went silent. Finally, someone from the back spoke up. I couldn't see who it was, but I heard him loud and clear.

"We're going to need more evidence before we change forestry practices."

I couldn't count the number of times I had heard this in my career. That was science. Science did not stand up and shout, We must change this now! Science said, *Gather more evidence.* Almost every journal article said, *More research is needed.* Science was careful. Science was plodding and slow. We posed a question and we set out to answer it. We tested our hypothesis, then retested again and again. But no matter how measured and meticulous, science alone was not enough to solve our planet's environmental crises. You can love science—and I do, deeply—but I knew it was not enough.

If science *had* been enough, the repeated warnings from scientists over the decades would have succeeded in bringing about a profound change in human behavior. For decades, the scientific community had been sounding the alarm that a warming planet posed a threat to life on Earth. Eventually, they realized they needed to use blunter language. In 2014, the UN's Intergovernmental Panel on Climate Change released a report saying that unchecked greenhouse gas emissions would trigger *severe, pervasive and irreversible impacts.* In 2019, a group of eleven thousand scientists from 153 countries issued a statement saying the climate emergency would cause *untold human suffering*.[2] With those chilling words, science had done its part.

In the aftermath of the conference, I knew we needed to not just publish our findings, but engage with the public and advocate for policy changes. We needed to continuously support local people in adapting and developing regenerative methodologies and economies for their communities so they would have healthy, resilient forests as climate changed. And we had to keep collecting data to document what remained and what was being lost, and discover how these forests could be restored. So, Jean and I set about gathering more evidence.

SINCE OUR PROPOSAL to establish an additional research site in Haida Gwaii had been turned down, the Haida forests were not represented in our Mother Tree Project data. This had me worried. We expected that the carbon pools in the maritime forests and bogs of Haida Gwaii would be deeper than any of the Mother Tree research sites, and alarms about the potential impact of logging in this climatic region had already been rung by scientists in the United States. The moist marine forests of western North America, they urged, ought to be priorities for protection to buffer against ecosystem collapse with climate change. These were among the oldest, most structurally diverse of all temperate forests—rich in carbon and species, and at low risk of wildfire. I realized that we needed to get busy sampling the Pacific coastal forests.

That summer, with the help of our devoted crew of students, we ran plots up and down the shoreline, in hypermaritime, maritime, and submaritime climates; on midslopes, upper slopes, and lowlands; in new clearcuts and in old growth; in Kwakwaka'wakw, Heiltsuk, and Haida territories. One hundred plots and counting.

By the time we'd finished sampling up the coast, we'd encountered humus layers up to two meters deep and rotting logs three meters high. Some of the most abundant ancient buried carbon was located on Haida Gwaii, and the most carbon-dense organic soils were in the salmon-rich riparian forests, with soil pools reaching almost 550 metric tons of carbon. In the trees growing alongside salmon rivers, we'd

Teah Schacter, twenty-four, excavating a soil pit in an old-growth cedar-hemlock forest near Quatse Lake in Kwakiutl territory on northern Vancouver Island, 2024. Teah is collecting mineral soil at five intervals to one meter depth for determining soil carbon content. Each soil pit takes about three hours to sample following the National Forest Inventory methodology.

estimated they alone contained over 1,100 metric tons of carbon—off the charts. Combined, the total ecosystem carbon stocks in the salmon-rich forests of Haida Gwaii were three times that of our coastal Mother Tree forests at Malcolm Knapp. No wonder Lisa described the forests surrounding the salmon streams on Haida Gwaii as the heart of the archipelago.

Clearcutting, we found, eliminated 95 percent of the aboveground carbon stocks from the mature forests on Haida Gwaii, aligning with our findings elsewhere in the province. These now-mature second-growth forests were rich in soil carbon not just because the forests were extremely productive and hence targeted early by logging, but because loggers a century ago left massive logs behind. These logs slowly rotted into the soils, helping retain some of the largest soil carbon pools outside of the boreal forests. But the methods of logging had changed, and today, with modern machinery and more thorough removal of the trees, we found only two-thirds of total soil carbon remained following clearcutting.

Of course, with the carbon pools went the forest-dependent mosses and plants. Where there was 65 percent cover of mosses of twenty-two species in the old forests, less than 1 percent cover and half the species remained after the skidders had been through. The thirty species of lichens and liverworts were also reduced dramatically, by two-thirds.

WE WERE COLLECTING SAMPLES in a thousand-year-old cedar-hemlock forest on British Columbia's south coast, about 600 kilometers (370 miles) southeast of Haida Gwaii. I dug deep into the moist dark humus for carbon analysis, feeling more like a groundhog than ever.

"You're going to fall into that hole!" Jody Holmes teased. Jody was an advisor to the Kwiakah Nation, which had asked for help assessing the condition of their territory in the Phillips Arm watershed, and strategies for restoring their forests from industrial clearcutting.

Phillips Arm is the traditional territory of the Kwiakah people, a small nation whose members live mostly in Campbell River on Vancouver Island, about an hour and a half away by boat. After more than a century of colonial dispossession, the Kwiakah were returning to their territory and ancestral village sites, recovering sovereignty over their stolen land. Fierce advocates for healing their industrially impacted forests, soils, and waters through the practice of regenerative forestry, they were partnering with various research initiatives, including the Mother Tree Project. Jody had brought me on this reconnaissance along with a team that included Chief Steven Dick and his brother Mike Dick of the Kwiakah Nation and forest ecologist Rachel Holt. Mike carried a shotgun in case we startled a grizzly.

Phillips Arm was only 250 kilometers (155 miles) northwest as the crow flies from our Mother Tree Project site at Malcolm Knapp Research Forest. Since it was a similar maritime forest, we figured our data and historic maps and records could help us reconstruct what the old forests used to look like before the legacy of industrial logging had left its mark. This comparison could provide us with a baseline of old-growth features we might try to recover with our restoration activities.

The lowlands of Kwiakah territory had been thoroughly clearcut by the logging industry in the past 150 years, leaving tiny patches of old growth on the steep upper slopes. In our pursuit of soil samples in the remaining old growth, we had crawled along mossy granite cliffs, over van-sized rotting logs, and through jungles of plush salmonberry and gnarled salal to arrive at this plot. The earthy, petrichor smell of the ancient soil had us feeling giddy. Perched on the side of this steep slope, we'd discovered the forest floor was one meter deep—over five times thicker than any of our interior Mother Tree Project forests.

Along with our crew, Jody and I crossed a roaring river to reach a third-growth Sitka spruce plantation one kilometer up the Phillips Arm valley. The river had widened into a hundred-meter boulder field when a debris torrent let loose a few years back. To cross, we had to jump from boulder to boulder over the rushing water. The valley once filled with old-growth cedar-hemlock forests had been logged in the mid-1800s. Some of the second-growth stands were logged again in 1995, and replanted with spruce monocultures in 1996. We finally reached the thirty-year-old third-growth plantation, the trees so densely packed that I had to fling my body forward to break through the wall of prickly branches.

Jim Pojar, a government forest ecologist, botanist, and natural history author, once dubbed these plantations "cellulose cemeteries," given the elusiveness of any life surviving among the doghair saplings. The understory was too dark for vascular plants, mosses, or lichens to grow. Absent were the slender beaked moss and large leafy moss we'd seen in the old-growth patch earlier that day, as were the ferns and wintergreens. Rainfall could barely penetrate the crowns, leaving the ground dry and covered in bristly brown spruce needles. In the middle of the plantation, the blade of my shovel immediately hit mineral soil.

Jody and I looked at each other, stunned. The forest floor was only a tenth of a meter deep. I sampled the surface layer and found it rough with partially decayed needles and twigs, the underlying humus dry, chunky, and fibrous. I figured its carbon concentration would be much less than the silky, chocolaty loam of the old growth.

The data came back a few months later. I had been right: the plantation forest floor carbon concentration was only 20 percent, compared with 60 percent in the rich humus of the old growth. With the forest floor this shallow and the carbon concentration in it so low, our calculations revealed that this replanted clearcut had suffered an *80 percent loss* of forest floor carbon content compared with the old-growth forest.

This was an even greater loss than we'd found in the clearcuts of the Mother Tree Project, and my mind churned to understand what could have caused it. There was clearly more at play here than just the carbon loss due to the logging machinery. Then it hit me. With the great, vibrant trees gone, replaced by small saplings and an empty understory, little photosynthate was now flowing into the forest floor. While the slash from the logging would have primed the saprotrophs, any remaining roots, soil organisms, and fungal networks would have been impoverished. The bacteria, fungi, mites, spiders, and earthworms, once abundant in the carbon-rich soils, would have been increasingly diminished. The saprotrophic fungi, supplanting the mycorrhizas, would likely in time cause the organic layer to wither and the soil pores to shrink. There were no longer plants adding organic matter to the earth.

These forests had been among the most productive ecosystems in British Columbia, with the world depending on them as large carbon sinks, and they were performing far below their potential. The forest floor had all but collapsed.

9

PLACE MATTERS

Jean and I gathered the measuring tapes, calipers, clipboards, and cruiser vests from the equipment room in my basement. At 8:00 a.m. sharp, our crew of students arrived in their rattletrap vehicles and helped us load up the field trucks.

"Are you ready?" I asked Nava. She and Hannah were coming with us to Kootenay Lake to assess the impacts of the treatments on the ecosystems. Two years had passed since we'd planted and it was time to measure the seedlings and returning plant growth.

We were all happy to be headed back to the forest, eager to resume the odyssey of fieldwork during this strange time. It was the first summer of the COVID-19 pandemic and the whole country was on lockdown. We'd had to get special permission from the university to gather in a group. We had to follow special protocols, wear face masks, use copious amounts of hand sanitizer, travel only two to a vehicle, and limit time spent with anyone not in our pod.

As everyone packed the gear, I surveyed the state of our transportation—six vehicles in various states of fitness. We had two solid field trucks—nicknamed the Silver and Blue—while the rest were a mishmash of student beaters. We'd had to rely on the students bringing their own vehicles because there just weren't enough field trucks to accommodate the two-person COVID limit. Nava was driving Grampa Pete's Chevy Silverado, its floorboards still muddy from getting stuck at John Prince. Danica's old Subaru Forester was making funny knocking noises, and she wondered aloud what that orange check-engine light meant. Aidan and Jacob needed to gas up their 1980 Pontiac, its back seat piled with clothes, sleeping bags, and a guitar, the tape deck loaded with Pink Floyd. Jean and I exchanged a look. It was going to be tough getting the crew all around the province in this wagon train.

For the third time in the course of our experiment, we were evaluating the plant life—trees, seedlings, shrubs, herbs, and mosses—returning to each of the five overstory treatments in our nine forests. We'd evaluated them just before the harvesting treatment had been applied and had returned immediately afterward to measure the impacts of the logging. Now we were back, two years after the seedlings had been planted.

It rained our first day, a warm, steady downpour that soaked ball caps, makeshift rain gear, and field notes. It made the work—which was unfamiliar to most of the new students—especially challenging. Taking these measurements was grueling, but it was the only way to determine the answers to our questions. Would the responses of the trees, seedlings, and plants differ between the clearcutting and partial retention treatments? Would the understory communities diverge from the primary forests, our uncut controls? Would these effects

Jean Roach taking notes at John Prince Research Forest in Dakelh and Sekani territory, 2018. The trees are interior Douglas fir, and here they are at their northern limit in North America.

vary depending on the location and climatic conditions of the forest? Would retaining more overstory trees help protect understory seedlings and plants from heat and drought stress at the arid southern sites, or from heavy summer frosts at the cold northern sites?

After ten hours in the pouring rain, we'd finished measuring only half the eighty-one subplots in the first clearcut. Danica had started counting hemlock germinants, hundreds of them no bigger than roofing nails, and after an hour she looked at me, eyes wide, pleading for advice. While checking a metal tag, Teah tripped over a stump and was instantly soaked by a paper birch dumping water down the back of her neck. Nava spent half an hour searching through the dripping cedar saplings where she'd dropped her forest-green water bottle. Aidan and Jacob spent over an hour finishing their first plot, getting every sleeve, sock, and strand of hair sopping wet from brushing up against the thick growth of aspens, thimbleberries, and fireweed.

"We have to speed up," I told the crew the next morning, "or we'll be here until hell freezes over."

Jean watched, smiling. She'd supervised hundreds of students over the years and knew all the tricks to get things going, but she was letting me handle this. I snuck her a look.

"We have sixteen hundred and twenty subplots to do at Kootenay Lake, and we completed only forty yesterday. At this rate, it will take us two months to finish this site, and we have to be moving to Jaffray in two weeks," I said.

The students shuffled their feet, listening.

"We averaged four trees per person-hour yesterday," I said. "I know we have to do a lot at each tree—find it in the bushes, mark it on the map, measure its height and diameter, assess its health and vigor—but it shouldn't be taking twenty minutes per seedling."

I told the story of my old boss Ted who, when I was their age, instructed me to double my speed writing planting prescriptions. I'd been miffed. *Didn't he know how hard it was?* Turned out I could work pretty darn fast when I put my mind to it.

The crew doubled their speed to eight trees per hour that day. In a few more days, we'd be cruising at eighteen trees per person-hour.

When we finished measuring Kootenay Lake in a total of fourteen days, just under the wire, I praised the crew effusively. Once every-

Grampa Henry (right) and his brother Wilfred Simard, wrangling logs to prevent them from jamming in the swiftly moving Shuswap River near Kingfisher Creek, British Columbia, in Splatsin te Secwépemc territory, ca. 1930. The work could be deadly, with men sometimes crushed by bobbing stringers or swept away in the rapids and drowning. The trees were felled with crosscut saws and dragged by horses to handmade flumes, where they were then sluiced downhill to Mabel Lake. The logs would empty into the lake, where Grampa Henry would use his homegrown tugboat, *Putput*, to drive them to the mouth of the Shuswap River, where they were boomed overwinter. At spring breakup, the brothers would release the logs into the river to make the thirty-five-kilometer journey to sawmills downstream at Enderby and Sicamous, British Columbia.

Temperate interior Douglas fir forest in Nlaka'pamux te Secwépemc territory at Twaal Creek, 2018. The forest is at the hot, dry end of our climatic gradient in the Mother Tree Project, and lies above the arid grasslands near Spences Bridge, British Columbia. The big trees are over two hundred years old and have seeded in a mutigenerational age distribution and multilayered size structure. Our graduate students mapped the architecture of common mycorrhizal networks and conducted kin selection research in this forest type.

Train of western redcedar in Kwakwaka'wakw territory near Nimpkish Lake on Vancouver Island, 2021. Cedar is the Tree of Life to the Indigenous people of the north Pacific coast, providing bark for weaving hats, clothing and mats, and wood for making houses, canoes, masks, and poles. It holds immense cultural significance, particularly through its connections to the spirit world. Indigenous people ask the tree permission to harvest, and in an act of reciprocity, provide a gift such as tobacco or sweetgrass, and care for the habitat of the ancient trees. At the current rate of commercial forest harvesting in British Columbia, one decade of logging on Haida Gwaii alone would fill a bumper-to-bumper train of cars spanning the distance from Los Angeles to New York City.

Sitka spruce mother tree in Kwakwaka'wakw territory near Port Alice, British Columbia. Our research has found that the largest, oldest trees, the mother trees, are the most highly connected with compatible neighbors through belowground mycorrhizal networks.

Mother trees emerging from the clouds in Simpcw te Secwépemc territory at the confluence of Mud Creek and the North Thompson River near Blue River, British Columbia.

Devil's club growing in the understory of western redcedar in Sinixt, Ktunaxa, and Syilx territories at Kokanee Creek, British Columbia, 2025. Devil's club is a sacred medicinal plant to Indigenous people, who use it for healing lung and stomach ailments, arthritis, pain, diabetes, and coughs, and use the stems for making fishing lures and spears. It is also a deeply spiritual medicine, helpful in purification, luck, protection, strength, and healing. If the plant is treated disrespectfully, there is less chance it will help. Devil's club grows abundantly in the understory of old forests, but dwindles and dies when the forest is clearcut.

Grizzly prints in Gwawaenuk territory at Nimmo Bay, 2023. Grizzlies have curved claws up to ten centimeters long that come in handy for rolling rocks and digging for crabs and clams.

Tidal estuary in Kitasoo and Xai'xais territories near Klemtu, British Columbia. Klemtu was created when the two nations, who spoke different languages, were forced onto a single reserve. It is located 200 kilometers (124 miles) east of Haida Gwaii and 55 kilometers (34 miles) northwest of Bella Bella in the Great Bear Rainforest on the central coast. The Kitasoo and Xai'xais manage the forests, fish, and wildlife in their territories using their traditional knowledge systems.

Old growth typical of the inland rainforests of British Columbia, where western redcedar, Pacific yew, and Douglas maple grow in community. In Sinixt, Ktunaxa, and Syilx territories near Kokanee Park, British Columbia.

Amanda Asay checking the seal on the airtight bags covering seedlings in Eva Snyder's doctoral experiment in the University of British Columbia greenhouse on Musqueam territory, 2021. They are preparing to label the seedlings with carbon dioxide tagged with the carbon-13 isotope to trace its movement among western redcedar, Douglas maple, and western yew, and whether it flows through a shared arbuscular mycorrhizal network, the soil, or both belowground pathways.

Raising a monumental pole, a *gyaa'ang*, at Tlielang on the northern spit of Graham Island, Haida Gwaii, 2017. The six-hundred-year-old *ts'uu* western redcedar was carved by master carver Kihlguulans Kihl 'Yahda Christian White, eight apprentices, and three journeymen over six months. The carvings tell the story of connection among the people and their relations, including the grizzly, beaver, eagle, hawk, sparrow, frog, pine martin, clam, and raven and the supernatural world.

Lisa White-Kuuyang and Haida women at the *gyaa'ang* raising at Tlielang.

A Haida Gwaii island shorn of its forest, 2021.

Danica Law, twenty-three, on her first day assessing impacts of the seed tree treatment on seedling regeneration in Sinixt, Ktunaxa, and Syilx territories at Redfish Creek, 2021. Danica is holding a 3.99-meter plot cord that she's sweeping around a center point to delineate the 50-square-meter plot for assessing the species, size, and condition of each seedling. Her coworker, Teah Schacter, is standing beside the plot recording the data and making a map of the location of the seedlings in the plot. There was a steady foggy drizzle the first day on the job.

Soil pit excavated in the Cheakamus Community Forest in Squamish and Lil'wat territories near Whistler, British Columbia, 2024. The soil is a Ferro-Humic Podzol, where the organic forest floor (litter, fermentation, and humus layers) is underlain by a thin white Ae horizon, reddish Bf horizon, and less well-developed BC horizon. A twenty-centimeter square of forest floor is first sampled, then mineral soils collected in standard increments to one meter in depth. The forest floor is meticulously sampled with a scalpel to ensure no minerals are mixed with the organic material. This soil contained approximately 300 tonnes (660,000 pounds) of carbon per hectare.

Nava, aged sixteen, and Liam Jones, aged seventeen, at John Prince Research Forest in Dakelh and Sekani territory, 2018. They are laying out a thirty-meter transect for measuring coarse woody debris and depth of the forest floor. Each transect followed a straight line in a randomly selected direction, requiring careful forward- and back-sighting with compasses, and climbing over and ducking under fallen trees.

Alyssa Robinson, Mother Tree crew member and doctoral student, 2025. The trees are coastal Douglas fir and Sitka spruce, and the plants are sword fern and salmonberry.

Tree-sit at Ada'itsx (Fairy Creek) in Pacheedaht territory, 2021.

Angela Davidson (Rainbow Eyes) of the Da'naxda'xw Awaetlala Nation, speaking in defense of the forests at Ada'itsx (Fairy Creek) on the steps of the British Columbia legislature in Victoria, British Columbia, August 2021.

The Itsekin village site, Ma'amtagila territory, is distinguished by the bright white clam middens on the beach and the diversity of tended nut and fruit trees just upslope of the shoreline. The tree species can include apple, plum, cherry, maple, and hazelnut for food, and cedars, hemlocks, and spruces for medicine and materials. The diverse village site is a garden and is distinct from the surrounding coniferous forest, characterized by a more monotonous texture where patches have been clearcut and replanted. The Ma'amtagila people have never ceded their land and in recent years have been returning to their villages and territory to restore and regenerate their forests.

Makwala, Rande Cook, hereditary chief of the Ma'amtagila Nation, created the *Dance of the Fungal Kingdom* in honor of the people working together like a vast mycelial network to renew and restore the lands and waters of Ma'amtagila territory. The dance was performed at the Mungo Martin House, Victoria, British Columbia, in 2025, by dancer Myumi Willie, daughter of hereditary chief Mike Willie of the Kwikwasut'inuxw Haxwa'mis people of the Kwakwaka'wakw Nation.

thing was stowed and ready to go, Jean decided to quiz the students about their observations. These tailgate sessions were lively and fun, and everyone gathered around, eager to participate.

"Have you noticed any patterns in the data?"

Ari pulled out their phone to make some rough calculations. In all of the treatments where old trees were left behind, they'd seen masses of naturally regenerated seedlings—especially Douglas fir.

"Sixteen thousand seedlings per hectare were growing in the sixty percent retention, but only two thousand in the clearcut," they said, jotting down the numbers.

The only seedlings in the middle of the clearcut were those we'd planted the previous years. No others had regenerated from seed naturally dispersed from old trees at the clearcut edges. Of the planted stock, seedlings we'd purposefully migrated from warmer climates and seedlings grown from local seed had survived best. The migrants from southern climates suffered from some frost damage because they had come from locations with less seasonal frost and therefore hadn't been pre-adapted to the new conditions. But they hadn't suffered too much. They also grew as well in the 30 percent overstory retention as in the clearcuts. The canopies of the old trees were protecting these migrants from the frost and drought that was more evident in the clearcuts. The good survival of migrant seedlings was encouraging—it meant that we could transfer seed from southern climates with reasonable success as the climate warmed. And retaining a partial canopy of old trees gave them cover from the elements. For now, at least.

In all the plots where overstory trees were left behind, whether 10, 30, or 60 percent of them, naturally dispersed seeds—which we called "naturals"—were germinating like the dickens. Not just Douglas fir seedlings, but also western redcedar, grand fir, western hemlock, paper birch, black cottonwood, and trembling aspen. The more overstory trees we had left behind, the greater the density of natural regeneration.

"It looks like there's plenty of naturals as long as there's an overstory tree nearby," Kaya agreed, looking over Ari's shoulder at their notes. Ari estimated that leaving overstory trees every ten to twenty meters, as we'd done in our seed tree treatments, would shed enough seed to fully stock the forest with naturals alone. Company foresters would be happy to learn they could just leave behind mature trees to promote natural regeneration, instead of clearcutting and having to plant new trees. Or at least they could augment their plantation seedlings with the naturally regenerating seedlings.

But there were no natural germinants in the clearcut treatments, where the timber edge was sixty or more meters away.

"Hey, this is interesting." Danica was thumbing through the stack of data sheets in her clipboard and noticed that more hemlocks, cedars, and grand firs were coming back in the 60 percent retention treatment. She'd learned in her classes that these species prefer shade, but seeing them here in the thicker areas of the forest would embed this knowledge deep in her bones. "It makes sense because they photosynthesize well in low light."

She was becoming an excellent field ecologist, I thought.

There were more western larch, paper birch, and trembling aspen in the clearcut and seed tree treatments. The broadleaf trees were sprouting and suckering because they loved the greater light and soil water in the more open conditions. But everyone had noticed that all of the native tree species were readily establishing in the partial retention treatments, as long as there were mature trees no farther away than sixty meters.

The students knew that the richness of the regenerating tree species would promote ecosystem productivity. They understood that synergies among plants, collaborating, competing, and interacting with one another, enhanced resource availability for the whole community. They'd learned these concepts in their undergraduate forestry degrees and could see them playing out right before their eyes here in the forest. Trees, plants, and animals living and working together in a diverse community.

Gaelin, who wanted to be a wildlife ecologist, thought the shade-tolerant trees coming back in the small gaps might start attracting more old-growth-dependent understory plants, birds, and mammals to the stand. He was starting to understand why clearcuts favored deer, who prefer browsing the tender shoots of new growth. Yet at the same time, the clearcuts were pushing the southern mountain caribou to extinction, as their primary food source was the arboreal hair lichens, horsehair and witch's hair, that grow only in primary forests.

"There's invasive plants in the clearcut and larger gaps, too," Nava said, picking bull thistle spines and red raspberry thorns out of her fingers. Her T-shirt was coated with dandelion parachutes, woollyhead wings, and grass seeds—the same grasses we saw in the lawns in Nelson. The grasses that had occupied the mineral soil exposed by logging were spreading rapidly. The main offenders were couch grass, seeded for erosion control along roadways, and tufted English ryegrass, used for lawns and golf courses. These tenacious species had spread into the forest from the roads and dwellings lining Kootenay Lake and were infiltrating the woods as the logging roads were punched farther up the valleys. Our data suggested that these invasive species were displacing some of the native plants. The aggressive grasses could also be changing the soil biology and hindering natural regeneration of trees.

Checking her plant lists, Eva noted that we'd lost a bunch of forest-dependent species from the clearcut and seed tree treatments. She flipped through the data sheets Jean had organized in a binder over the past couple of weeks. There were fewer native shrubs and herbs, and almost no mosses and lichens where the forest was heavily cut.

The clearcut was especially distinctive in how it had transformed the plant community. The yews had disappeared with the overstory cedars and maples. The shade-tolerant pink wintergreens and fairy slipper orchids were gone, and the forest-loving false azaleas and oak ferns had diminished. The luscious, forest-dependent mosses—step moss, Oregon beaked moss, and the leafy mosses—were scarce in the clearcut and seed tree treatments, and the pelt and coral lichens had disappeared following the clearcutting. If any lichens had escaped the skidders, they'd been too sensitive to survive in the open. None of the arboreal lichens remained.

Bree, a graduate student who would later analyze these plant communities in depth, worried that logging was deeply affecting the cryptogams—seedless plants like ferns and mosses. Despite their diminutive size, mosses and lichens sequester one-third of the world's terrestrial carbon. Cryptogams form a spongy protective layer—called a cryptogamic crust—that acts as a physical barrier, protecting existing carbon pools by regulating water tables and soil temperature. Bree's master's thesis work would confirm what we had all observed in the coastal ecosystems: that the mosses and lichens were the most vulnerable to clearcutting.

"The industry loves clearcutting, but they could make a pile in their planting program if they left seed trees, and it would be less destructive to the plant community and soils," Jean said. But she knew the companies would argue that the savings from not planting wouldn't offset the extra time and money it took to leave some trees behind.

At the end of summer, we would publish our Kootenay Lake results.[1] Jean crunched the numbers, the students compared our findings to the literature, and I composed the first draft. With the COVID lockdown restricting our movements, we met regularly on Zoom to discuss the results. The unlogged forests had the most diverse plant community and the deepest carbon pools, hands down.

But the reality was that clearcutting was not slowing in British Columbia. Quite the opposite: immense pressure was building to log forests, either to reduce fire risk or to get the wood out ahead of conservation efforts. Logging companies were rushing to log the old trees before they could be set aside into permanent preserves. Our study suggested that leaving seed trees standing every ten or twenty meters, or even better, aggregated in patches interspersed with openings no more than two tree lengths in width, would result in an optimal balance of seed source proximity, mycorrhizal inoculum potential, and resource availability. Retaining trees in this configuration ensured the most diverse tree regeneration and protected seedlings, plant biodiversity, and carbon stocks.

We all wondered what we would find at our other experimental sites, and whether the results would differ according to the climatic conditions of each place.

OUR NEXT STOP was the most arid forest, Jaffray. After a couple weeks there, we would follow the gradient toward the cooler forests northward.

When we arrived at the head of the logging road at Jaffray, the sky was azure with a few popcorn clouds. We were in a hot dry spell and wildfire hazard was inching up. Everyone had brought their swimsuits for a dip in glacial Lake Koocanusa at the end of the day.

"Call every kilometer," I instructed the drivers of the Silver and Blue, our only two vehicles with two-way radios. Hannah was driving Blue, so I was especially anxious. "Warn the logging trucks there are three vehicles without radios. Everyone else, stay close behind."

Hannah, with Eva riding shotgun, headed the pack along the dusty gravel road. Nava and I followed in Dad's Chevy Silverado, then Danica and Aidan in their beaters, and Liam and Jean brought up the rear in the Silver. The rest of our crew were belted into the back seats.

"Empty pickup at one kilometer, three more trucks without radios," Hannah radioed to any oncoming vehicles. I could see her making the call through her rear window.

Within two kilometers, a fully loaded logging truck—three meters wide, twenty-five meters long, and weighing 40,000 kilograms (88,000 pounds)—appeared over a rise, accompanied by a tornado of dust backlit by the early morning sun. Driving down the dead middle of the road, the truck barreled toward Hannah, who kept on, steering slightly to the right. Hannah was expecting the logging truck to shift left. It didn't. I pulled to the side, clutching my steering wheel.

"Pull over, pull over, pull over," I muttered, raising my voice with each word as if Hannah could hear me. The truck kept coming.

"Fuck! Get into the ditch!" I shouted.

Just as the truck bore down on her, Hannah swung Blue into the ditch. I was speechless. Nava sat paralyzed. Quickly, I checked my rearview mirror. Our train of vehicles had all driven into the ditch. We all sat idling as we absorbed what had just happened.

Hannah rolled Blue back onto the road. Slowly, we all followed and continued on.

Twenty minutes later, we arrived at the first replicate of our Jaffray site. I gave Hannah a hug without letting on that I'd almost had a heart attack. I told her she did great and reminded her to always pull aside a few hundred meters before the trucks are expected to pass by. I could tell the experience had shaken her.

Jaffray was in the western foothills of the Rockies, the air dry as a bone after passing eastward from Kootenay Lake over the Purcell Range, dumping rain and snow on the high mountains. The rainforests of cedar and hemlock in the Selkirks had given way to interior Douglas fir, lodgepole pine, and western larch in the rolling foothills of the Rocky Mountain Trench, with ponderosa pine woodlands in the valley bottoms.

I was excited to see how the forest was responding to the overstory treatments. I rifled through my gear, found my cruiser vest, hat, and water bottle. Plot cord, measuring table, calipers. Four pencils. *Where are my boots?* After pulling out the entire back of the pickup, I sat on the ground. I had forgotten my work boots back in Nelson.

"Give this a try, HH," Jean said, handing me a roll of purple duct tape.

I wrapped my Birkenstocks with enough tape to make them function like boots, then looked around, ignoring Jean's sniggers at my purple-taped footwear. The seed tree treatment was to our right, the mature Douglas firs beautifully spaced about twenty meters apart. The control was to the left, hundred-year-old Douglas fir forest with pinegrass, soopolallie shrubs, roses, and heart-leaved arnica in the understory. The 60 percent beside it was a little overcut, but still within the range of what we needed. Across the road was the clearcut, looking hot and dry. Beyond it, the 30 percent, the clumps and gaps each about twenty meters wide.

We worked in crews of two, each pair moving in parallel up the lines of subplots. Some of us started in the clearcut, others in the seed tree.

My purple Birks stirring up dust, I soon realized the clearcut had become a dead zone for seedlings. Only 5 percent of the Douglas fir seedlings we'd planted had survived. They'd either shriveled from the heat, succumbed to spring frost, or been browsed to a nub by white-tailed deer and Rocky Mountain elk. The plant community couldn't help protect these little trees in the extreme conditions. The twinflowers, red-stemmed feathermosses, and reindeer cup lichens that prevented desiccation of the soil in the forest had been replaced by thistles and orchard grass brought in by ranchers and their cattle. These offered little shade and had transformed the soil food web into a parched, hostile environment for the ectomycorrhizal conifer seedlings. Some invasive mosses, with names like purple horn-tooth and bristly haircap, that tolerate extreme temperatures in places like roadsides and mine tailings had encroached on the mineral soil exposed by the logging equipment. These were just doing the basics of gluing together the mineral particles rather than actually helping the trees.

By noon, the heat was so intense we huddled in the shade of the pickups, dousing ourselves in water.

"Take three liters of water each to the field," Jean said, checking the crew for signs of heat exhaustion.

Loaded with water, Jean set off with Ari and Eva to finish the seed tree treatment, while I went with the rest of the crew to complete the 60 percent overstory retention treatment. We got into the rhythm of the measuring and found 33 percent of the seedlings we'd planted in the 60 had survived. Just as good, the seed tree and 30 percent patch retention came in at 32 percent survival. The seedlings migrated from warmer climates survived as well as the local seed, all benefiting from the protection of the overstory trees. Seedlings that had been moved from northern forests had mostly died in the heat.

"*Peltigera aphthosa!*" Danica called out from the 60, and we rejoiced that the pelt lichens had survived the selective harvesting, as had red-stemmed feathermoss, broom forkmoss, and yellow salsify. There were even some electrified cat's-tail moss and reindeer cup lichen nestled among the trees, and yarrow, rosy pussytoes, and common juniper on the mounds. A woodpecker hammered at a snag, which were now absent from the clearcut and seed tree treatments.

The pink sunset was upon us on our last day at Jaffray. Where we'd been drenched in rain at humid Kootenay Lake, we'd been baked in the sun at arid Jaffray, and the trees and plant communities were telling us two different stories. As we sat outside the Travelers Inn, Jean prompted the crew for the patterns they'd seen over the past two weeks.

"The sixty percent retention treatment is the best," Jacob said. His face was still swollen from the seven stings he'd sustained after stepping on a wasp nest. His pal Aidan had also been stung trying to help him escape. I winced in sympathy, and Jean went over to apply thimbleberry poultices.

The 60, where the majority of old trees had been left standing and the understory trees thinned from below, had lost only 8 percent of its carbon stocks, a stark contrast to 50 percent of the carbon disappearing from the clearcut. The seedlings were doing well in the 60, and so was the plant community and cryptogams. Alexia's wildlife cameras, which she'd installed for her master's thesis research, had captured white-tailed deer, Rocky Mountain elk, coyote, wolves, and cougar in the 60 percent retention, almost as much wildlife as she'd seen in the unlogged controls. The canopy provided cover and the understory plants provided forage. The elk seemed not to mind the clearcuts for the increased forage they provided and they found safety from predators like wolves in the nearby forest patches. Removing the ingrowth of understory firs, whose densities had dramatically increased with fire suppression, had also reduced fire risk.

JOHN PRINCE RESEARCH FOREST in central British Columbia was the last of our field sites. We finished our work there just as the aspens were turning yellow and the snow was settling over the mountains north of Tezzeron Lake.

"Watson, I do believe we're done," I said to Eva, as we finished our last plot.

"Brilliant, sir!" Eva said.

Groaning, the crew threw their plot cords at us. As we headed down the road toward the trucks, I felt hopeful running under the arbor of arms slapping high fives.

We sat in the cookhouse that night scarfing massive piles of nachos and reflecting on our results. Our backs were sore and our minds were numb, but I could only feel gratitude at the crew's devotion to our experiment and the pursuit of this knowledge. The steady progression of our discoveries reminded me of an eagle soaring higher and higher, allowing us to see and understand greater and greater complexity in the forest. Our findings were clear: clearcutting at the northern limit of Douglas fir had created harsh conditions for reforestation. With the tall trees gone, the water table had risen and thick brush fields of Canada goldenrod, horsetails, and red raspberries had choked the seedlings. Frost from radiative cooling and cold air flows had killed thousands of seedlings. The more overstory trees retained, though, the warmer the understory remained at night and the greater the diversity of regenerating seedlings and native plants.

When we examined the migrated seedlings closely, we discovered seeds originating from warmer climates survived and grew best under 30 to 60 percent overstory retention. These treatments, where the canopies of the overstory trees buffered the understory trees and plants against severe frosts and droughts, offered a variety of conditions for the establishment of planted trees and seed sources for natural regeneration. The naturals stood alongside the newly planted migrants, and their co-establishment could be a crucial step in helping the forests adapt to the changing climate. The plant community was also more balanced, better reflecting the composition of the old forest.

As we drove away from John Prince the next morning, I spotted a mama grizzly, probably the same one we'd seen a couple years earlier, and her two cubs ambling down the road, having escaped the fires this year. They looked fat and ready for winter. The clearcutting had destroyed their dens, but our controls had left them intact, ensuring the bears had homes to hibernate in through the cold season.

The students, Jean, and I had measured nearly one hundred thousand seedlings over the course of two months—an enormous accomplishment, and the data would be full of treasures. It was already clear that the location of a forest mattered deeply to the way we tended it. Each of our nine forests was unique, and the data would tell a different story from one regional climate to the next. In contrast to the homogeneous application of clearcutting practices regardless of the ecosystem type, our data clearly showed that *place mattered.* We needed forest practices that were site specific. Instead of one-size-fits-all clearcutting, the harvesting, restoration, and regeneration methods needed to be crafted according to the specific attributes of the ecosystems and the people who lived in and cared for them. This would provide more work for the local people and create healthier connections to the land.

When I returned home to Nelson at the end of August, my mind was full of possibility. Each forest along our gradient had a unique climate, soil, and species composition, but this variety was what made our craft so energizing and creative. It also provided diversity and resilience to this crucial forest landscape. The forests needed specific tending in their special places, and there was no single recipe for all. Their unique connections would need protecting and their adaptive cycles would require careful attention as droughts lengthened, heat waves became more severe, and rains became more erratic as climate changed.

As we had expected, regeneration of forests and protection of plant biodiversity in harsh climates, whether due to intense heat or extreme cold, benefited from retention of overstory trees when stands were harvested. The trees protected the understory plants from desiccation and frost damage and provided seed and nurturing for the next generation of trees. *All* of the seedlings and plant communities flourished best when there was a mature tree overhead and no more than ten or twenty meters away. The old trees provide protection from the elements, mycorrhizal networks to tap into, and a healthy soil food web to deliver nutrients and water. Even under the most favorable conditions, elder trees were still needed to nourish the regenerating communities.

Again and again, our data was leading us to the same conclusion: with their time-worn genes and deep-rooted connections, the old trees were essential to the rebirth of the forest. The forest was a multigenerational place where young trees flourished in the protective understory of the mother trees, and the mother trees transmitted their wisdom and knowledge to the young, ensuring life for future generations.

My thoughts went to Teresa and Dave's conversation in Spences Bridge. They were correct that we needed to leave enough trees so they could connect and communicate with one another and protect the animals, birds, and food and medicine plants, and the soil, water, and air. These were Indigenous values that applied to the whole world. No longer could we afford to sacrifice these life-giving principles for the profit of a few. Forest policies needed to change so that enough trees would be left behind in each specific place to protect ecosystems and people.

The teachings were age-old and crystal clear. As Lisa had said, "We must have respect for all living things. That is a responsibility we all share."

10

RECIPROCITY

That spring, Amanda drove eight hours over the still-icy mountain passes to Vancouver while Eva crossed the choppy waters of Georgia Strait from her parents' house on Vancouver Island. Working in batches of thirty pots a day, with the help of an undergrad assistant, it took them four days to complete the isotopic labeling of 120 pots of seedlings in the UBC greenhouse.

This greenhouse study was Eva's first big labeling experiment. It mirrored her Septuplets study in the field, but instead of planting seven seedlings in the ground, she'd planted one each of cedar, maple, and yew together in pots. In some pots, the roots of the three trees were separated by a very fine mesh that permitted only water exchange. In others, a coarser mesh was used, allowing the mycorrhizas to penetrate and connect the seedlings. In a third treatment, Eva used no mesh and allowed the roots and mycorrhizas to fully intermingle.

She had spent a year prepping and planning for this moment, growing and tending her seedlings. Would photosynthetic carbon travel among the three species? How much would move through the arbuscular mycorrhizal networks compared with the soil? Would shading increase the flow of carbon from cedar to yew and possibly influence yew's paclitaxel production?

I was on campus that week, and I stopped by the greenhouse after class one day to check in. Watching them work, I marveled at how attuned Amanda was to Eva's desire to prove her own competency. Amanda had become an expert at labeling trees during her doctoral studies. Whenever possible, she stood back and let Eva take the lead while keeping a careful eye on things and making the occasional joke. It reminded me of the times I'd seen her play ice hockey. As team captain, Amanda played defense, supporting her teammates, passing the puck when needed, helping other players score goals. That was Amanda's whole way of being—always ready with a smile and an offer to lend a hand. I was grateful that Eva was benefiting from Amanda's mentorship. Just as my postdoc Brian had done when he'd helped Amanda label older Douglas fir with carbon-13 to trace its movement to nearby kin, Amanda was now passing the wand to Eva.

The two of them placed the gas-tight plastic bags over Eva's seedlings, which were healthy and fully leafed out. Amanda demonstrated how to place putty around the stem of each tree to seal the bag before injecting isotopically labeled carbon dioxide into the atmosphere of the seedlings.

Syringe in hand, Eva paused before depressing the plunger.

"Does it matter if you do this slowly or not?"

"You don't want to go too fast," Amanda said, "but it doesn't need to be super slow." Eva cautiously pushed down the piston. "I think you're going at a good pace."

It had been raining earlier and was overcast, but as the afternoon lengthened, the sun came out and shone brilliantly through the greenhouse windows, heating up the notoriously damp old building. We all peeled off our layers. Eva rolled up her shirtsleeves and Amanda shed her Whistler sweatshirt.

When the last seedling had been labeled, Eva sighed in relief at having completed this crucial step in her experiment.

"Amanda," she said, "you are my rock."

Now working as an ecologist for the Ministry of Forests, Amanda was settling into her life, moving her work forward and playing sports at a championship level. She lived just a few blocks from me in Nelson, though we'd both been so busy since she graduated that we hadn't had much time to visit. As always, Amanda was running at top speed. She'd bought a hundred-year-old home where she'd planted a vegetable garden, and she was working three jobs—researching carbon pools in the inland rainforests around Nelson, publishing her doctoral thesis papers, and playing baseball, now in her sixteenth year with the women's national team. A few months earlier, she had told me she wanted to continue her own research on kin selection in the mountain forests surrounding Nelson. She wanted to understand how the relatedness of trees in the rainforests enhanced their natural regenerative capacity—how kinship among trees might help increase the resilience of these ancient places against climate shock. She had mentioned her idea, testing my thoughts.

"Do you think the ministry would support such experimental work?" she wondered.

"They might." Especially if she could show that encouraging natural regeneration would cost less than planting, and that retaining old trees would also keep the forest understory cool and reduce fire risk.

THE SUMMER OF 2021—the year of the West Coast heat dome—was shaping up to be the hottest on record, and drought levels were off the charts. It was so hot that, earlier in July, 619 people had died from heat-related causes in Vancouver. Mum and Kitty had been confined to her living room, Mum opening the doors and windows at night and running fans during the day. That day, I was checking on Eva's Septuplets experiment at the Kootenay Lake site, where she'd planted seedlings in the wagon-wheel pattern: cedar in the center, surrounded by yew, maple, and Douglas fir. But the forest hadn't seen rain for a couple of months and we worried that the young trees wouldn't survive the brutal summer without supplemental water. I packed my truck with jugs of water to bring to Kootenay Lake, while Eva did the same at her northern sites.

When I got to Eva's experiment, I could immediately see that the Septuplets were flagging. The maple leaves were drooping and some of the Douglas fir needles had turned red. I coughed as I hauled a dozen four-liter jugs from my truck to water the little trees. The lightning that came with the heat had ignited 1,610 wildfires in British Columbia that year and filled the valleys with orange smoke. Six fires were burning in the mountains surrounding Nelson, and my eyes were stinging. As I emptied the jugs onto the ground one by one, the water seeped instantly into the earth, the soil gasping for more.

The Septuplets watered, I walked into the forest across the road to sit in the shade before heading back to Mum's house. It was refreshing and cool under the old cedars, and I was calmed by their incense. The doctors had reviewed Mum's application for MAID, evaluated her medical charts, and interviewed her. This was no impulsive decision but the result of much thought and hours of discussion with her doctors, family, and friends. The process was painstaking, and now Mum was waiting to learn if she qualified. I shook my head at the thought. Mum might be ready for the next phase of her life cycle, but I wasn't. While I supported her right to chart her own course and die with dignity, I just couldn't accept that her time was up. Sure, she was forgetful, but to me, she still seemed so youthful, so vibrant. During her last stint in the hospital, stricken by a *Clostridioides difficile* infection, she'd made the interns giggle when she put the chamber pot on her head. She still voted and wrote letters to elected officials, still insisted on attending local protests to oppose the government's feeble policy on old-growth forests.

In late May, Robyn's husband, Bill, and I had taken Mum to a demonstration at the Ministry of Forests in Castlegar, where about a hundred people were protesting the clearcutting of old growth and the police action against our kids at the Fairy Creek blockades on Vancouver Island. Nurses, forestry workers, grandparents, mothers, and children walked alongside one another, carrying signs with slogans broadcasting their concern for the future.

"Save our old growth! Save our old growth!" Mum chanted, savoring every minute of the protest, her COVID face mask dangling beneath her chin. She wore a sign around her neck that Bill had made for her, reading, "Old people for old growth."

In the square in front of Minister Katrine Conroy's office, several young people lay on the sidewalk, their signs reading, "Respect existence or we die." One man held a chainsaw, pretending to fire it up and cut people down like trees. Mum didn't understand why they were lying in the middle of the road.

"It's Extinction Rebellion," I whispered.

Mum turned to talk with a forester whose sign read, "Save our water, save Box Mountain." Mum had grown up in Nakusp, just around the corner from Box Mountain. Her best friend, Lorraine, had lived in an old wartime house across the hayfield from the cedars and hemlocks on the northern slope. Lorraine had died from breast cancer in her midforties, and Mum had never really gotten over the loss.

"What's happening at Box Mountain?" she asked, her eyes welling up.

The forester explained that the old-growth trees of Box Mountain were on the auction block in the name of wildfire mitigation. The town council had teamed up with a local timber company and gotten a license to clearcut the mountain slope under the rationale that it would reduce fire risk to the community. I had walked the forest with the local folks and found it old, diverse, wet, and at little risk of burning. Mum had taken us kids swimming at Box Lake, just below the mountain, many summers back in the 1960s. I remembered the old forests casting cool shade over the mountain lake. The community wasn't even close to the mountain.

"Save Box Mountain!" Mum chanted, beaming at the forester.

The minister never did emerge from her office that day to explain her government's decisions on old growth. None of the government's actions made sense to me. We urgently needed a province-wide moratorium on logging ancient forests, not temporary deferrals, vague promises, and empty rhetoric about protecting trees—a strategy Canadians called "talk and log." Instead of slowing down, the rate of clearcutting old-growth forests in the province was speeding up.

THE SUN LIT UP the thin threads of lichen linking the branches of the old cedars to the young ones. Scattered among the old and young trees were the dead and dying. Snags and candelabras, conks and widowmakers emerged from luscious crowns of hemlock boughs and cedar braids. Midcanopy beaked hazelnuts, Douglas maples, and bitter cherries stretched elegantly through the gaps created by the departed, capturing the sifting light. Their changing leaves cast an orangey-pink hue over the seedlings emerging in the sun dapples on the forest floor. The connections were here, right in front of me. Death was just one phase in the cycle of life.

Mum had always been ahead of the curve. She'd been the first in our family to earn her master's degree, and she'd bravely decided to get a divorce in the 1970s, when it was far from commonplace in our

Mum, eighty-five, protesting against the logging of old-growth forests in Sinixt, Ktunaxa, and Syilx territories at the Ministry of Forests office in Castlegar, British Columbia, 2021. The demonstration was held in solidarity with the land defenders at Fairy Creek.

community. I'd always been proud of her guts and tenacity. Every weekend when I was in high school, she'd led a gang of kids zooming down the Headwall and Cariboo black-diamond ski runs at Tod Mountain, catching the first and last chairlifts of the day. She'd skied into her early eighties, finally gravitating to easier runs. She'd learned to windsurf in her fifties and had spent many glorious days sailing the lakes of Nitinat and the Nicola, drinking beer at the pub at the end of the day. She was still ahead of the curve now, insisting on planning her own exit strategy. Intending to stay free until the end, that was simply Mum being Mum.

Change was hard to take sometimes, and I just wanted time to slow down. A western larch towered above, its departing needles hinting yellow, a dying western white pine standing nearby. The hemlocks and firs added a sharp resinous tone to the sweet cedars, reminding me this was about Mum, not me. I sat on a log and pulled out the apple I'd picked from Mum's tree. Using my knife to carve out the worm, I bit into its tart flesh.

MY FIRST BOOK had been published that spring, and the reception had been better than I'd dared to imagine. The book had made the *New York Times* bestseller list and been widely and glowingly reviewed. *The New York Times Magazine* had published a beautiful profile of my work. I'd been interviewed and profiled in dozens of national and international magazines and newspapers, and requests for speaking engagements were flooding in. A Hollywood production company had even bought the film rights. Sitting alone in the woods eating an apple, I had a hard time believing all of this was really happening. While my TED Talk in 2016 had made me something of an internet celebrity, the book had garnered a level of recognition I had never dreamed of. The trees had given me a voice, a platform. If my newfound renown was a little overwhelming, it was also hugely rewarding to know that my work was helping to awaken people to the intrinsic value of the forest. I knew that many scientists never got this opportunity, and I wanted to make the most of it in service to protecting these sacred places.

The cedar and yew saplings crowded around me with hushed arms, so I pulled at them gently and found them anchored to the soil. Both species had been layered in here by their parents. The leaves of the younger ones were absorbing the orange and red part of the light spectrum that had filtered through the overhead layers, and captured them for photosynthesis, and any carbon subsidies through the belowground networks would have been heartily welcomed in the shade.

Burrowed under the conifer skirts were huckleberries, oak ferns, and feathermosses, capturing far-red light even farther along the photosynthetic spectrum. The orchids, wintergreens, paintbrushes, and lilies absorbed the weakest shards of light, and some relied in part on heterotrophic carbon delivered through fungi and haustoria, those feeding apparatuses attached to taller neighbors. Lichens, mosses, and liverworts, living in carpets on the forest floor and in the crevices of tree bark, were adept at using the dimmest of sun flecks. This complex structure enabled the forest to capture and cycle the full spectrum of photosynthetic light filtering through the forest canopy all the way to the forest floor.

Seedlings of hemlock were germinating on my rotting log, so I plucked one to examine its diminutive roots and mycorrhizal fungal threads. White and pink filaments flowed from the fine root tips, and I imagined them drinking water from the spongy wood and feeding on nutrients. These germinants were linked into the mycorrhizal networks of the elder hemlocks nearby and receiving carbon subsidies in the deep shade. This process, important in the regeneration of old-growth western hemlock forests, had been discovered by my student Gabriel Orrego, who had defended his master's thesis in 2018, the year before Amanda defended hers.[1]

I pushed my trowel into the crunchy forest floor and dug deep. The soil was disturbingly bone-dry at the surface but still moist a foot deep. My training in soil ecology had given me night vision into the underground world, and I'd learned from decades of study that the complexity of underground biological communities reflects the diversity aboveground. This enables individuals, species, and guilds with specific rooting characteristics, mycorrhizal symbionts, microbiomes, and metabolic pathways to access soil nutrients and water in different niches of the soil profiles. Douglas fir elders root deeply, compared with younger saplings, and pull water from unplumbed caverns. Their mycorrhizal fungal partners of older successional stages access water in pores that simpler, less robust fungi on younger trees cannot. The older trees circulate some of the water to seedlings in shallower soils via their mycorrhizal root systems.

The more niches the community of species fill, the more fully the resources in the ecosystem are absorbed. This process—the building and filling of different spaces in an ecosystem—is called niche complementarity. Cottonwoods, aspens, and willows are especially good, compared with other tree species, at accessing soil phosphorus with their dual arbuscular and ectomycorrhizal fungal symbionts, and autumn litter fall—leaves, stems, bark, and other organic matter—redistributes this essential element to others. The cedars and maples are particularly adept at tapping into calcium and magnesium pools, some of the minerals located in deep layers and others at surface horizons, which they redistribute to licorice fern, maidenhair fern, and spiny wood fern through stem flow and litter turnover. Each tree species has different skill sets, accessing remote patches of nutrients, twisted into caverns of dissolved elements, and rich pools of organomineral complexes in different horizons of the soil. These complementary strategies and synergies not only access the greater profile of resources but enrich the whole resource pool, benefiting the entire forest.

Some species also seem to make something out of nothing. Sitka alders in the intermediate layer of the forest, large-leaved lupines in the treefall gaps, and lung lichens on tree branches can fix atmospheric nitrogen into forms that can be absorbed into their tissues. There is no way, other than biological nitrogen fixation, rainfall, or incoming streamflow, to add usable forms of nitrogen to the forest. In the fall, this nitrogen in the alders, lupines, and lichens would be distributed to other plants through litter fall and decay. The hemlocks can better access this dispersed nitrogen on aerated knolls, while cedars can absorb it in the waterlogged dips, resulting in more complete use of the fixed nitrogen in the whole ecosystem.

The mixing of trees—species, genotypes, sizes, and ages—enhances not just their own productivity, but the diversity of birds, mammals, insects, and fungi, too. Innumerable bats, birds, and beetles, and even much larger creatures like bears, live in the hidden furrows and dens of trees, logs, and soils. All of these species, in their own way, scatter the seeds, fruits, nuts, and spores, and rely on the forest to provide a variety of habitats. A diversity of trees, in filling and creating a range of niches, provides even more food sources and nesting sites. The short natural lifespan of the birches, aspens, alders, and cottonwoods, which generally live less than a century, creates snags where insects live, mushrooms bloom, and woodpeckers excavate cavities. The longer-lived cedars and hemlocks provide a steady supply of food and habitat. The birds and mammals living in this multifaceted township of trees facilitate broader fertilization, pollination, and germination. The interactions and connections among the many creatures of the forest are sophisticated and self-organizing, creating a complex system that is continually changing, adapting, and cycling.

Just as in a community of people, diversity makes the forests more resilient and resistant to disturbance, and this enhances and stabilizes carbon sequestration and storage. With greater tree species richness, age ranges, or even genetic variability within a single species, there are fewer fires, floods, and infestations that will compromise long-term carbon pools. The presence of paper birch neighbors helps reduce infestation of firs by bark beetles or root pathogens by enhancing soil nutrients—and hence tree vigor and defense—in the otherwise acidic conifer soils. Deep-rooted conifers plumb subsurface water and redistribute it through the ecosystem. Broadleaf trees interwoven with conifers also pump water from below and reduce fire risk because their succulent leaves dampen the flames.

Finishing my apple, I tucked the core into a hole, noticing the roots of a nearby birch wrapped around the roots of its Douglas fir neighbor. If only the essence of the forest—diverse, connected, reciprocal, cycling—could be embedded in forest policy and management practices. In my home province, as in many forests around the world, forestry practices have historically focused on establishing single-species plantations, although this appears to be changing. Diversity of species, structure or age heterogeneity, and gap function have been purposely designed right out of the forest through clearcutting, monoculture planting, weeding out native plants, and harvesting on short rotations. Old mother trees, which fill crucial niches and contain the rich genetic diversity of response to previous climates and disturbance regimes, are usually hastily removed because their great size makes them so profitable.

These practices are based on the misconception that productivity in the forest results primarily from competition, rather than the sophisticated system of niche complementarity and interspecies collaboration as well. Deprived of species and structural diversity, the monoculture plantations are unable to capture the full spectrum of light, water, or nutrients in an ecosystem. They also miss opportunities for synergies, the chance to make more from less in communities with multiple species.

Nonetheless, a group of scientists recently published a paper clearly demonstrating that aboveground carbon stocks in plantations around the world are 70 percent greater where there are up to four species planted compared with one.[2] Data from the Mother Tree Project's nine experimental forests agrees with the meta-analysis—that above- and belowground carbon stocks naturally increase in our forests as tree species richness climbs from three to nine species.

I thought back to the stonewalling from foresters in the 1990s when I'd presented research showing that plant diversity helped conifer plantations. Back then, studies documenting enhanced productivity in mixed stands were rare compared with the plethora of articles expounding the virtues of single-species stands. At least now, the science was more widely accepted. I hoped that our data, along with our findings about the importance of protecting the mother trees, would help influence policymakers to change forest management practices.

GIVEN THE DYNAMIC, cyclical nature of the forest, I considered it a gamble that Eva would find overlap in the arbuscular mycorrhizal fungal species among cedar, yew, and maple in the inland rainforests of British Columbia. But she did find it, loud and clear. She also discovered that carbon flowed underground between the three species. Her results from the Septuplets experiment and her complementary greenhouse experiment suggested that these three co-occurring species were likely woven together in a belowground mycorrhizal fungal network. And Eva's finding that overstory composition also influenced the belowground linkages between the understory yews, maples, and cedars confirmed our hunch that *each place matters* to the nature and intensity of the connections.

The little forests Eva had planted into pots were attuned to one another, adjusting carbon pools by titrating the fluxes among them. Cedar and maple were shuttling photosynthetic carbon to yew, and the diminutive medicine trees were returning small amounts to their neighbors. The more shade cast on yew, the more carbon cedar and maple transmitted to the little medicinal tree. The larger trees appeared to be transmitting photosynthetic energy that could serve as a vital supplement to yew's low rates of photosynthesis deep in the shade, and I wondered if yew could be reciprocating with cedar and maple through the transmission of small carbon-rich defense molecules. These neighbors of different species appeared to be cooperating and competing with one another and benefiting the whole. The three species working together would enhance the total productivity of their community, allowing the trees to capture more fully the different wavelengths of light filtering through the canopy, the various compounds of nitrogen distributed among the soil pores, and the water held with assorted strengths in the variegated layers of soil.

Eva and Amanda's discoveries built on my own doctoral work, in which paper birch and Douglas fir transmitted carbon isotopes back and forth through belowground pathways, the amounts increasing with the amount of shade birch cast on the understory Douglas fir. Cedar nearby received small amounts of isotope from its early successional neighbors, and this may have given the tree a small boost to persist and eventually grow to old age. In these forests, birches and firs were early successional species that in a century would be overshadowed by the old-growth western redcedar.

Eva and Amanda's findings, together with my research, suggest that trees in forests have powerful abilities of perception and communication, and they coexist in a finely attuned dance of reciprocity. These intelligent relationships can provide guidance for our own behaviors, not only among ourselves as human beings, but with all of our relations. This knowledge deepened my conviction that we needed to protect the remaining intact forest ecosystems, not only for their diversity and life force, but for their wise teachings.

II

THE FOREST DEFENDERS

"I feel like a bug," Nava whispered.

I knew what she meant. *Overwhelmed and insignificant.* I squeezed her hand as we stood in the midst of a sea of sun-bleached stumps. Forty football fields of ancient forest had already been cleared from the north side of the Fairy Creek drainage (or Ada'itsx, the name belonging to the Pacheedaht Nation, in whose territory it resides). We picked our way slowly through the clearcut, climbing over broken limbs and severed trees. Yellow cedar and mountain hemlock had been cut calf height, the wood and bark on the stump faces ragged from the saws. I felt nauseous as I clambered over bones of branches jutting from the soil. My chest was heavy and my legs like cement. Mineral soil lay bare and forest floor was left torn asunder where cables had dragged logs from the corners of the clearcut to the common landing areas. I saw no sign of fairy lichens or pipecleaner mosses, no fairy slipper orchids or fairybells. The plant community and all the creatures in symbiosis with it had been torn up, crushed, or buried under the slash. This was a war zone, and few living things had survived the assault.

It was the summer of 2021, and a battle was raging between the industrial logging company Teal-Jones Group and a cluster of activists who had been camped out for months in an effort to protect Fairy Creek's headwater forests from being logged in the coming days. In news articles about Fairy Creek, the activists were labeled radicals and ecoterrorists. For their part, they called themselves forest defenders, and they were putting their bodies on the line to protect the ancient trees from the saw. Fairy Creek was one of the most pristine watersheds on Vancouver Island, and the only watershed, outside of parks and reserves, left untouched by industrial forestry on the southern end of this island the size of Belgium. The centuries-old trees provided habitat for critically endangered migratory birds and other old-growth-dependent species.

The alarm had been sounded last August by a teenage climate activist named Joshua Wright. Home on COVID lockdown in a tiny community on Washington's Olympic Peninsula, Joshua used satellite imagery and geospatial data sets to monitor logging activity and identify old-growth forests at risk of destruction. He was scouring satellite photos on his computer when he noticed the telltale scars of a new logging road blasted into the side of a mountain. Teal-Jones, he learned, was harvesting stands of ancient cedar around Fairy Creek. By alerting fellow activists on the ground, he ignited the movement my daughter and I had come to Ridge Camp to support.

Nava looked bewildered as I explained that the clearcutting followed provincial law. This practice was legal according to the conditions of Tree Farm License 46, held by Teal-Jones Group, and was certified by international standards as sustainable. The road we had followed up the San Juan Valley had enabled the company to clearcut nearly all of the lucrative timber around Port Renfrew. It was part of a massive logging road system on Vancouver Island that was harvesting old-growth forests throughout the entire Timber Harvesting Land Base—forest land owned by the Crown. With each forest logged, there were irreversible losses in carbon and biodiversity. The people defending these trees knew the models: it would take hundreds of years to recover the carbon stocks, while the world had less than a decade to act.

A rugged, sunburned young man offered his hand to us in greeting, smiling through his thick beard.

"You can call me Cake," he said. Most of the forest defenders were using camp names—pseudonyms—to protect their anonymity.

Cake led us quickly up the steep grade to greet Moe, a mother of two with curly dark hair tucked under a blue scarf, who'd taken leave from her business building sustainable housing to join the protests. We hopped over a sleeping dragon, a pipe cemented into a hole in the ground to which the activists would lock their arms to prevent police from hauling them away. Moe's fingers were raw from piling rocks and building hard blocks around these devices.

I could hear the urgency in Moe's voice as she briefed us. The police might raid Ridge Camp because of a protest planned at the British Columbia legislature. She couldn't afford to get arrested because she needed to get home to her kids. I was slated to speak at the legislature with Pacheedaht elder Bill Jones and other chiefs and activists, and we needed our message to be clear and level-headed, to bring the issue to the public's attention without further aggravating the police.

We visited several defenders at the cookfire, sharing breakfast, strategy, a smoke. Their camp names were Gadget, Cog, Beard. Fireweed, Huckleberry, Lichen. They were scientists, teachers, doctors, nurses, students. Carpenters, riggers, and climbers. Settler-rooted folks and Indigenous people. One was an artist, another a professor of wildlife biology. All were full of equal parts anxiety and resolve.

Today they needed to fortify a five-meter-tall tripod tree-sit: three fir poles lashed at the top, with a hammock hanging down for the sitter. Six tree-sits were scattered along this ridge, some accessible only with climbing ropes. Gadget and Cog worked on the ropes over coffee, trading tips on how to pee from high in a tree. The tripod was situated directly over the centerline flagged for bulldozers, which were ready to push the road over the ridge and into the headwaters. Nava and I had seen them from below, the yellow machines poised in the lower cutblocks.

A woman called Rainbow Eyes, of the Da'naxda'xw Awaetlala Nation, had volunteered as a tree sitter. Her face lit up when she saw us. She thanked us for coming, saying our support was crucial in this fight. Rainbow Eyes, as with many at Ridge Camp, was part of the Rainforest Flying Squad, and she and Bill Jones would become prominent voices of the Fairy Creek movement.

Moe led us to the commanding tree they called Titania. Standing tall at the headwaters, estimated to be two thousand years old, Titania was the elder mother tree in this amphitheater of old growth. Her branches were draped with antique mosses, lichens, and liverworts. Marbled murrelets nested in her crown. These secretive seabirds nest only in the upper limbs of the largest old-growth trees, sometimes flying up to eighty kilometers inland in search of ancient conifers with huge moss-coated branches that will cushion the single egg they lay each year.

I figured Titania was connected through her fungal networks to her neighboring sister cedars, yews, and ferns, as suggested by Eva's research and that of my previously graduated doctoral student, Shannon Guichon. The understory species would likely receive carbon or nutrient subsidies from her in the deep shade, and the nests in her crown would be shared by a web of birds and mammals. The integrity of the Fairy Creek drainage—its hydrology, biodiversity, and carbon pools—seemed to depend on her. The red ribbons marking the proposed road into the Fairy Creek watershed ran directly through Titania. Without the defense of the tree sitter, she would be gone.

Nava sat close to me in the protection of the giant tree's roots. We all agreed that the protest was deeply complex. Teal-Jones Group was reaping the benefits of logging the old growth, but so was the local Pacheedaht Nation. And not just the Pacheedaht—many coastal First Nations were in a devil's bargain with the colonial government. Their cultures had evolved with these old-growth forests, yet impoverishment, land dispossession, and the upheaval of their traditional governance systems by colonialism were compelling them to log their own homes.

At Fairy Creek, the Pacheedaht elected council had a revenue-sharing agreement with Teal-Jones Group and had repeatedly asked the activists to leave and to respect the sovereignty of the nation. The elected chief and council were uneasy about the increasing acrimony in the fight over old-growth logging. They were also responsible for providing social services to their nation while paying down debt to the provincial government, which had offered them a loan to run a sawmill specializing in milling old growth. The nation needed to log Fairy Creek to keep from defaulting on their loans.

But elder Bill Jones questioned the authority of the elected council members, saying they didn't represent the will of the nation or its hereditary governance system. The elected chief and council, he said, were an illegitimate result of Canada's colonial Indian Act.[1]

The activists, both settler and Indigenous, viewed themselves as the trees' last line of defense. They knew the old trees were essential for a livable future, for clean air and water. They understood that their best strategy was to slow down or disrupt the logging operations while allies elsewhere lobbied the government for policy that would halt the clearcutting altogether.

The next day, I spoke in defense of old growth on the steps of the legislature, along with Bill Jones, hereditary chiefs David Knox of the Kwakiutl Nation and Rande Cook of the Ma'amtagila Nation, and many others. We chastised the provincial government and emphasized the strong public sentiment: 82 percent of Canadians were solidly against old-growth logging. It was the one-year anniversary of the Fairy Creek movement, and hundreds of people had gathered in support, holding up colorful hand-painted signs about protecting old-growth forests. Aunties, uncles, grandparents, and children from the Kwakwaka'wakw and Nuu-chah-nulth Nations danced in regalia to the drums and voices, each song telling a story of the interdependence of the people and the forest. Without the old forests, their ancestral livelihoods were at risk. I felt nervous when it was my time to speak. So much was at stake, and I wondered whether my message would even be heard by the powerful people holding the pen on forests. But I stepped up to the mic and called out the government for failing to protect these ancient places.

That same morning, the Royal Canadian Mounted Police conducted a raid of Ridge and Waterfall Camps. As RCMP drones and helicopters circled overhead, armed officers in tactical gear dismantled the barricades, destroyed the encampments, and arrested dozens of forest defenders, many of them Indigenous youth.

IN OCTOBER, I returned to Fairy Creek with Lynda Mapes, a reporter with *The Seattle Times*. Fairy Creek had become the site of a massive and well-organized demonstration—the largest act of civil disobedience in Canada's history. At the height of the blockade, up to two thousand people were embedded in the woods in a sort of guerrilla insurgency. Tens of millions of dollars were being spent on police enforcement. More than eleven hundred people had been arrested for defying court orders and four hundred would eventually be charged.

Lynda had been reporting on old-growth forests for most of her career. She knew that people in the United States were paying close attention to what was happening in Canada's forests. With photographer Steve Ringman in tow, Lynda and I traveled up the Cayuse Creek watershed, on the other side of the Fairy Creek drainage relative to Ridge Camp. The logging had already started in the highest reaches of Cayuse Creek, and the RCMP were heavily armed and making arrests. We passed three police checkpoints where Lynda had to show her press badge. The easy smile she flashed for the Mounties concealed a veteran journalist's skill and experience.

Our crew passed sagging tents perched along rotten logs, sleeping bags drying in sunflecks and clothes airing in the mist. Faded T-shirts hung from a makeshift clothesline. We eventually found our way to the top switchbacks, trees already being felled to make brand-new clearcuts. Six young defenders emerged like crows from tree-sits and sleeping dragons at the end of the newly blasted road beneath the ridge separating the Cayuse and Fairy Creek headwaters. They wanted to show us a stand of ancient yellow cedars flagged for logging, with western screech-owls and marbled murrelets nesting in the canopy. Ushering us over massive logs and deep ravines, the defenders pointed out rare largeflower fairybells, western rattlesnake root, and little prince's pine along the way. A defender called Cookie showed me lungworts and pelt lichens. Another defender pointed out the rare old-growth specklebelly lichen draping the smooth bark of Pacific silver firs. This watershed was home to 326 species of animals, birds, plants, and cryptogams and untold numbers of fungi and microbes. Seventeen of these species were at risk of extinction.

No matter how small or seemingly insignificant, each species occupies a unique niche in the rich community of vertebrates and invertebrates that helps deliver water and nutrients to the trees. Lichens absorb water and fix scarce nitrogen from the atmosphere—processes that are crucial in maintaining the high productivity of this rainforest, with its massive carbon, nutrient, and water cycles.

"How are they allowed to log with so many endangered species present?" someone wondered.

I explained that the provinces governed the harvesting on forest land in Canada, and that provincial law superseded federal law when it came to resource development. The federal *Species at Risk*

Act was too weak to protect the habitat of endangered species in areas approved for logging.

We reached the grove of giant yellow cedars, including the center matriarch with a hollowed-out trunk. With a chilly wind at our backs, we peered into the tree cavity, and seeing it empty, squeezed through the tiny entrance and crowded inside. It felt like a cocoon, dry and with a soft floor of bark strips. Glimpsing coarse hairs snagged on decaying woody fibers, I realized a mother bear had molded this bed when she'd come to hibernate and give birth to her cubs. Black bears prefer to den inside big old-growth trees like this one. The large cavities keep the newborn cubs dry during their vulnerable first months and the shell of sapwood protects them from the bone-chilling cold and blowing snow in the depths of winter. The mother bear would stay up to six months, from late October to early May, awakening from hibernation in late January to deliver her cubs. She would return to this den each fall, fattened on berries and salmon. I picked one of her hairs from a velvety grain, the scent of wet grass merging with the citrus heartwood and fresh Pacific rain.

Huddled inside the den, the defenders worried about the impending fall rains and quietly hoped for early snow to delay the felling.

"Do you hear thumping?" I asked.

"Helicopters," someone whispered. We emerged to see a chopper churn over the ridge. Under the swirl of rotor blades, the cedar matriarch stood at sixty meters, encircled by her offspring. This grove had survived millennia of climatic fluctuations and windstorms and been fed by centuries of salmon runs. Defenses evolved over millions of years had helped the trees survive and accumulate as much carbon as an Amazon rainforest (measured in the surrounding forests at thirteen hundred metric tons per hectare). But these defenses were no match for the chainsaws.

I ran, slipping and skidding, with Lynda, Steve, and our companions across the steep slope to where the helicopter was hovering over a helipad constructed in the middle of a new clearcut being harvested. One of the defenders climbed up the platform and waved his arms to try to scare the chopper off. The rest of us dropped behind fallen logs, covering our ears. The collision of worldviews resounded like a thunderclap. It was part of the larger conflict over forestry in British

Columbia that dated back three decades, to the War in the Woods at Clayoquot Sound. Activism and alliances had helped to save the old-growth trees at Clayoquot and gave rise to the agreements creating the Great Bear Rainforest.[2]

As swiftly as it appeared, the chopper turned and flew back over the ridge, likely radioing our location to law enforcement. We climbed up the clearcut to the road, scrambling over giant fallen trees and airborne bridges. The specklebelly lichens were drying on the lifeless bark and lung lichens were littering the ground, no longer alive to provide crucial nitrogen. The young forest defenders were already on the gravel when I crawled onto the roadbed. Our police escort was waiting.

As I walked toward the truck, I noticed the chocolate-brown humus layer revealed by the road cut—over one meter thick, rich with carbon. Our research in forests along the Pacific coast suggested that at least half of the soil carbon in this forest was stored in this soil organic horizon. If our observations at Phillips Arm applied here, then once logged, most of this forest floor would be lost. After being bladed, churned, decomposed, burned, leached, eroded, and starved of photosynthate, anything remaining would be largely depleted of living carbon.

As the policeman pushed Cookie into the cab, he muttered that he was saving her from her own stupidity. I knew better than to argue. But as we rolled down the mountain, I couldn't help but try to explain.

"It takes decades for clearcut forests to stop emitting more carbon than they sequester, and centuries more to recover the sink strength of the original stands. We don't have decades for these forests to recover from clearcutting. In the hundred years it takes for a forest to mature, our planet could warm upwards of five degrees Celsius," I said. But he wasn't listening.

Weeks later, late at night, one of the defenders emailed me a video. I clicked on the link and stared at my computer screen. It was the ancient cedar matriarch and her sisters. The saws were deafening as they chewed through their heartwood. The wood cracked, crowns fell, branches broke. I covered my eyes.

I imagined the dying sounds. First the gasping of the roots for food. Then the starvation of the mycorrhizal fungi. The withering and the cracking of the filaments holding the soil particles together. Then the dripping of last rhizodeposits—the secretions, exudates, mucilages, volatiles, lysates, polysaccharides, glycoproteins, metabolites—as they trickled, for the last time, from the dying roots and fungi.

Then the crackling static. This was the sound of the microbes dying in the hyphosphere. And the last wisps of volatiles pulsing to wind. Without microbial biofilms to feed on, the soil food web—the protozoas, mites, collembola, nematodes—broke down. Then their predators—the millipedes, centipedes, beetles, and spiders—starved.

Then came the collapsing sound. The soil food web, diminished, could no longer hold the soil particles together. The aggregates and clods fell apart into dry, individual particles.

I put my hands over my ears as I imagined the forest floor falling. A great *whumpfing* noise rushed through my eardrums, like a building imploding.

The rains came next, gathering the unbound humus in its rivulets. The trickles turned into roars and the forest floor eroded into streams, filled the channels, and deposited in estuaries.

The mother bear would return to the location of the old matriarch with her belly full of cubs, finding no den. The marbled murrelet would come back to her nest bringing a meal of Pacific herring, finding her chick dead. The salmon would return upstream to spawn, to find their beds choked with humus. The defenders who had fallen in love with the great trees would carry their broken hearts away with them.

Nava returned briefly to the blockades with her friends that fall, hearing Titania calling her back. She avoided arrest, but some others were not as lucky. As the fall rains set in, the camps were dismantled and the sentences handed down. While hundreds of prosecutions for criminal contempt collapsed as courts found that the police had not properly read the injunction to protesters, many young people performed community service, served jail time, paid fines.

Rainbow Eyes was arrested five times and faced seven criminal charges, jail time, curfews, community service, and house arrest. Still, she told reporters outside the courthouse, "There is no price too high to protect our Mother Earth." At her sentencing, she argued that intergenerational trauma and the loss of Indigenous culture should be

Suzanne, Nava, and Titania, the yellow cedar matriarch, about two thousand years old, in the headwaters of Ada'itsx (Fairy Creek) in Pacheedaht territory, 2021. The logging company had flagged the location for a new road straight through Titania. Just over the crest behind us, the headwaters of the adjacent watershed had been clearcut.

mitigating factors in her sentence. I remembered her sparkling eyes, her devotion to the land and to natural law, and I prayed it would sustain her.

Relations among the provincial government, First Nations, the public, and the police were deeply convoluted. The courts ordered injunctions against the protesters to allow the logging to continue in the valleys surrounding Fairy Creek, and these would eventually open

Confrontation between police and forest defenders at Ada'itsx (Fairy Creek), Pacheedaht territory, 2021.

Police line facing forest defenders, Ada'itsx (Fairy Creek), Pacheedaht territory, 2021.

A forest defender arrested at Ada'itsx (Fairy Creek), Pacheedaht territory, 2021.

the door for Teal-Jones to resume clearcutting in Cayuse Creek. In its prosecution, the government lawyers disparaged the defenders and questioned the credentials of supporters. But the judge chastised the Royal Canadian Mounted Police for abusive behavior and for imposing restrictions on the media and members of the public that clearly violated their civil liberties. Justice Douglas Thompson wrote that the forest protectors were "respectful, intelligent, and peaceable by nature. They are good citizens in the important sense that they care intensely about the common good."

The amount of money spent on the police action against the youth and elders of British Columbia made no sense to the young people in court. Why not use the tens of millions to protect the forest and the communities dependent on it? Why not use the $24 million estimated earnings from harvesting the timber to develop value-added industries instead of continuing resource extraction? The Ancient Forest Alliance, working with economists at the environmental consulting company ESSA, did the calculations and found indeed that the forest left standing was worth far more than the wood products derived from cutting the timber.[3] When the carbon storage, tourism, recreation, nontimber forest products, coho salmon habitat, educational opportunities, and real estate values were taken into account, Fairy Creek's intact forests were valued at $97 million (in 2018 Canadian dollars, net present value).

In June 2021, the provincial government agreed to pause logging in the Fairy Creek area—including the disputed Tree Farm License 46—for two years while the Pacheedaht, Ditidaht, and Huu-ay-aht First Nations collaborated on a forest stewardship plan. In a statement called the *Hišuk ma c̓awak* Declaration, the nations said they were tired of watching outsiders make decisions about their Hahoulthee, or traditional territories, and that they would manage their lands in accordance with their own laws and traditions.[4]

The Rainforest Flying Squad and other blockaders welcomed the news, but said the agreement didn't go far enough to ensure protection for the entire Fairy Creek rainforest. With the support of Bill Jones, they said they would continue to stand their ground and defend the trees.

Garry Merkel, a member of the Tahltan Nation and co-chair of the independent review on the status of old-growth forests in British Columbia, said he didn't expect tensions over British Columbia's logging practices to cool down anytime soon. "We're going to have Fairy Creeks happen all the time," he said.

In 2023, the government of British Columbia would tack on another two years to its initial two-year logging deferment, offering the nations more time to complete their long-term management plan. Yet nowhere was permanent protection for the old-growth forests in the Timber Harvesting Land Base guaranteed. Teal-Jones Group went on to sue the Province of British Columbia and the Haida Gwaii Management Council in 2023 for $75 million—compensation, the corporation declared, for profits lost when it was forced to stop logging old-growth forests on Haida Gwaii.

THE FIRST ECOLOGISTS

A growing number of studies have shown that Indigenous-held land contained some of the most biodiverse and carbon-rich ecosystems in the world. The First Peoples know how to tend forests to protect and enhance their natural gifts, increasing ecosystem resilience and supporting global cycles. I hoped Western science could learn from their example.[1]

I had also seen the consequences of 150 years of dispossession. When Indigenous people were expelled from their territories, they'd been prohibited from carrying out their ancestral obligations as stream keepers, fire keepers, salmon tenders—responsibilities that had been passed from generation to generation. With the ancient relationships severed, the cycles of life were breaking down. The result was degraded forests, plummeting fish stocks, and rivers choked with logging debris. The forest had become so out of balance that predator populations in many places were collapsing. The Kwiakah Nation had even brought in farmed fish to feed starving wolves and grizzlies.[2] While working with my students to help restore biodiversity and berry plants to the industrial conifer plantations in Kwiakah territory, on the south-central coast of British Columbia, we had found the skeleton of a black bear killed and eaten by a grizzly, its flesh stripped to the bone. The scarcity of food had driven nutritionally stressed bears to kill and eat other bears.

After seeing my TED Talk on how trees communicate with one another, Makwala, also known as Rande Cook, one of the hereditary chiefs of the Ma'amtagila Nation, had reached out and invited me to visit the nation's territory. Rande saw that the way I understood the forest—as a connected, whole ecosystem—was aligned with his people's worldview. To Rande, the mother tree was the heart of the forest, the birthplace of life. He had taken me to a bear den inside an ancient cedar tree, to show me how the red soil inside the mother tree was the birthing ground for the mother bear.

The Ma'amtagila is one of eighteen nations that make up the Kwakwaka'wakw peoples. The nation had been forced from their territory onto four small reserves by the federal government in 1886. Their traditional governance and sociocultural systems had been outlawed and their children required to attend residential schools at Alert Bay or Port Alberni. But the Ma'amtagila had never ceded their territory to the government of Canada, and because no Ma'amtagila had been present for any treaty negotiations, Rande's nation was seeking formal recognition of inextinguishable rights and title in the courts.

Rande had asked me to help him document the effects of industrial clearcutting on his territory and to assist them in creating a plan to protect the remaining old-growth and primary forests, restore the damaged forests, and ban clearcutting. He hoped that our Western scientific methodologies, interwoven with the values, goals, and presence of the Ma'amtagila Nation on their land, could be helpful in making their case in the courts. Our work also resonated with him, and with Indigenous knowledge systems, because it sought to understand and protect the relationships, connections, and cycles of the forest. Using appropriate language to help prove the Ma'amtagila connection to the land—past, present, and future—was key in their fight for recognition and sovereignty in the federal courts. With legal status would come recognition of existing rights and title, and full access again to their land and resources.

The fjords, islands, and estuaries of the southern Great Bear Rainforest, between Vancouver Island and the mainland coast of British Columbia, are accessible only by water, so Mike Willie, a member of the Kwikwasut'inuxw Haxwa'mis Nation, would take us in his boat. Teresa and forest ecologist and environmental analyst Rachel Holt joined us to start the work of documenting and planning, and K'odi

Hereditary chief Makwala, also known as Rande Cook, Ma'amtagila Nation, 2021, inside a black bear den in an ancient western redcedar tree in Ma'amtagila territory. To Rande, the den's red earthen floor symbolizes birth and fertility, and he honors this connection in his art, dance, and song. This patch of cedars, one hectare in size, is surrounded by clearcuts.

Nelson, a renowned Ma'amtagila artist and songkeeper who had founded the language and culture camp Nawalakw where we were to stay, helped lead our expedition. Dozens of young people were now spending their summers at Nawalakw, reconnecting to their heritage, learning to speak their language, building trails, and tending the forests and rivers in a cultural immersive experience.

Our trip from Alert Bay to Hada, deep in Kwakwaka'wakw territory, would take a full day. As we sped across the sound in Mike's flat-bottomed aluminum skiff, the magnificent mountain fjords of Kwakwaka'wakw territory unfolded before us. Knife-edge arêtes spawned mist that flowed steeply down basaltic faces. Waterfalls flowed and water dripped steadily from rock ledges. Ancient Sitka spruces swung from crevices in rock cliffs. On the knolls and ridges, candelabras of ancient cedars spiked to forty meters. Trees of great ages leaned and hung in swamps and benches. On south faces, Douglas fir roots dug deeply into rock crevices. Where the earth was too steep, the soil had let go, leaving clean stone glistening on recent rockfalls and young hemlocks dangling in midair. Glacial air rolled down these mountains like thunder. A grizzly turned over stones on a beach with his great paws as he hunted for barnacles and chitons.

Our group spent the night at the Nawalakw lodge, and the next morning, with a dozen teenagers in six canoes, we paddled across the inlet from Hada. When we hauled our canoes up onto the rocky driftwood shore, Matthew Ambers, the camp leader and language keeper, sang with his drum in Kwak'wala, the language spoken here for thousands of years, asking the forest for permission to enter. I had learned the importance of this ceremony, which was repeated many times during this visit and my other visits to Indigenous-held territories. The ritual of asking permission demonstrated respect for the land and humility and gratitude for the waters, forests, and wild animals.

As the last drumbeat echoed off the mountain, Matthew nodded. The forest had permitted us passage.

The young people ran along the trail, teasing one another and playing tag among the trees. Their voices hushed as we approached a grove of great cedars, giants about fifty meters tall and each a couple of meters in girth. Matthew again sang and drummed for the trees, while Rande urged the kids to "listen—really listen." As the kids calmed, the sounds of the forest were unmasked. A raven gurgled overhead. A bald eagle whistled. The half dozen sister cedars had long since died, but looked as though they'd lived for thousands of years. Before departing, the ancient trees had spawned new generations of forest, layering fresh saplings, cohort after cohort, into the soil. In the process of dying, the trees had fueled a continuous cycle of life.

"We are Indigenous people of this land, and we are here to protect these forests," Rande told the kids. "In our language, we call this forest *nawalakw*—supernatural."

Rande explained how the trees were connected to one another and to the Kwakwaka'wakw people. How the old cedars and the people had lived in community and cycled together for many lifespans. How the forest and human societies were one.

"We have long known these trees live spiritually, in a society," he said. "We are interdependent with the trees. When they breathe out, we breathe in."

Echoing Teresa, he said, "Our people have tended, harvested, and carved these trees for thousands of years." Rande, who was a master carver with an intimate knowledge of the wood and its essence, shared the wisdom of how to live and thrive together.

"The land is our teacher and healer. The teachings of the ancient sister cedars, the grandmother trees, are to respect, love, and help one another. To hold each other up."

I stood among the *nawalakw*, the supernatural trees, the grandmothers. Their teachings, to watch over one another, stirred in the breeze and the breath shared among us.

On Rande's invitation, I described how we had used molecular biology and isotopic chemistry to trace the connections between older and younger Douglas fir trees, like the old and young cedars in this forest. The older ones exchanged photosynthetic energy and nutrients through their root and fungal systems with the younger ones, and the younger ones were supported by the old as they came of age. I explained how scientific tools had helped us understand that these trees could recognize and tend to their kin. That the genes of the trees had been shaped by their elders. That the trees were in collaborative, kincentric relationships—members of an extended ecological family that includes all living things. These scientific findings were closely aligned with Kwakwaka'wakw knowledge.

The kids were shy, but one asked if the young cedars were communicating with the elder sisters, and I said our research suggested so. In her second doctoral experiment, which built on Amanda's research with Douglas fir, Eva would investigate kin recognition among sister cedars by quantifying their growth responses and exchange of isotopic carbon through belowground networks.

In August 2024, a group of Kwakwaka'wakw hereditary chiefs organized a Sa l a, or mourning ceremony, to grieve the loss of the trees, nonhuman relations, and their traditional way of life due to clearcutting on Kwakiutl ancestral lands, on north Vancouver Island. Matt Ambers, the camp leader at Nawalakw, was the designated chiefs' speaker at the ceremony. "Every tree had a use, every tree had a spirit—that's all gone now," he said.

The next day, we traveled by motorboat southward from the Hada estuary to the village site of Itsekin, where Rande's people had lived just a century earlier. The village had been largely abandoned after a smallpox outbreak in the 1880s. A rich palette welcomed us as we approached. Sparkling white clam middens shone from the beach and headland soils. Crabapples, choke cherries, and plums planted around remnant house beds stood laden with fruit, and within steps were hazelnuts, highbush cranberries, thimbleberries, and huckleberries. In the lower canopy gaps were Saskatoon berries, elderberries, and salmonberries, and beneath these multiple layers grew wild sarsaparilla, Pacific coast strawberry, and wild ginger, all once carefully tended.

Behind the dwelling spaces, bigleaf maples and red alders were mixed among the cedars, forming a shady, fire-resistant shelterwood. Sections of bark had been stripped from some of the smaller cedars for weaving, and planks taken from larger living trees. The villagers had packed the niches of these gardens with complementary fruit, nut, and medicinal species. Even though Itsekin had been missing its people for decades, their careful stewardship over millennia was indelibly carved into the land.

The biodiversity in the village was much greater than in the surrounding forest, with lime, mint, emerald, and jade green brilliantly complex against the coniferous canopy. In the industrial forest, the understory was oppressed. But in the gardens, where plants and trees had been transplanted, coppiced, burned, weeded, fertilized, and pruned, the community of species was beautifully rich.

Not only did these tending practices enhance food diversity and production, they also reduced risks of drought, flooding, fire, and species invasions. The people had selected the village sites for food, water, and safety, and their stewardship enhanced these functional qualities, while improving habitat for bear, moose, deer, birds, and pollinating insects. Dark with organic matter, rich in nitrogen and phosphorus, the soils surely contained rich carbon and nutrient pools.

But such diverse forests were now rare. Other than the village sites and the postage-stamp-size patches of old growth perched high on cliffs, the landscape had been completely clearcut over the past 150 years. Our boat visited inlets and fjords, coves and straits, sounds and bays. Everywhere we went, the forest had been taken. Almost nothing was left untouched. I saw no groves of ancient cedars or hemlocks, no massive Sitka spruces along the shoreline. In place of wild forest, rows of Sitka spruce or Douglas fir seedlings had been planted to industrial standards.

The dense plantations of spruce or fir that stood here now had been sculpted to be as uniform as picket fences, with the native plants largely suppressed. The soils they grew on had been degraded by the crawler machines, leaving a cracker-thin forest floor. The forest had been cultivated with the notion that the second-growth stands would be as valuable as the old growth they were replacing. The trees had been planted too densely, to occupy the land with wall-to-wall wood. They were cellulose cemeteries, just like the third-growth plantation Jody and I had seen at Phillips Arm. The gray trunks of the replacement soldiers were covered in scabby branches. Niches that should have been occupied by shrubs, mosses, and lichens, amphibians and birds, were mainly vacant. The old mother trees that had once stood here, with bird's nests in their crowns and bear dens in their trunks, were gone.

Any plant considered a threat to conifer wood production had been snuffed out by the herbicide glyphosate, or Roundup. The dark, dense understories were empty of salmonberry, thimbleberry, and huckleberry. There was no devil's club or blue elderberry nourishing the forest floor, no maples, yews, or alders to rebuild the soil organic matter. A handful of mosses and lichens clung to flattened niches. The water and nutrients, inadequately absorbed, were washing out to sea. This was like a real-life version of Ursula K. Le Guin's sci-fi classic *The Word for World Is Forest,* in which men from a polluted Earth colonize another planet, clearcut its forests, and oppress its Native inhabitants.

Snaking through these clearcuts were roads built at maximum grade to get to as much timber as cheaply and quickly as possible.

Clearcuts in Kwiakah territory near Frederick Arm in the southern Great Bear Rainforest, 2022. Without permission from the Kwiakah people, the government of British Columbia gave lucrative licenses to the timber companies, which thoroughly clearcut the old forests. In some cases, the companies cut the same land twice. The Kwiakah people have repurchased the licenses on their own land and created the Mac̓inuxʷ Special Management Area to restore the forests and gardens using regenerative forestry and their traditional knowledge systems.

Second-growth stand after clearcutting old-growth forest. The regenerating stand is very dense, homogenous, and of low quality, with insufficient light for understory plant growth. The stump is all that remains of the old-growth cedar-hemlock forest that was cut thirty-eight years ago.

Seeding into the tracks were alders, madly fixing nitrogen to heal the raw earth. Slashes of billowing alder ran zigzagging like serpents over the land.

As the plantations in the clearcuts aged, and the smaller trees weakened in the shade of somewhat larger individuals, the picket fence stands would eventually naturally self-thin. With glimpses of sun warming the soil, seeds of shrubs and herbs buried for decades would try to germinate and bring the forest back to life. Still, the diversity of these forests was a fraction of their past glory, when hundreds of species lived connected among the ancient trees.

The earlier clearcuts, now over a century old, were in better condition. The returning forest was productive and diverse and the soil was rich and fertile. Back then, the timber along the fjords had been felled straight into the sea, then boomed and tugged to oceanside mills. Any roads that had been built were for horses or steam engines. By the grace of the hand fallers on foot, the land—the legacies of soil and seed—had been spared. The stands were left to regenerate naturally, beautifully.

The forest industry, by its own admission, preferred to cut more old growth because the industrial plantations that replaced them were worth only a quarter on the dollar. When the Mother Tree crew returned to sample the third-growth forest the following summer, our results were clear—the productivity and biodiversity of the planted forest were about one-quarter that of the old growth.[3]

A year later, a group of scientists armed with remote sensing, satellite imagery, and ground truthing would corroborate what our crew found on the west coast of British Columbia: most of the world's primeval forests have been degraded, and this is contributing to the Earth's climate crisis.[4] The global terrestrial carbon deficit—carbon once held in vegetation and soil but now released to the atmosphere through land-use change—is hovering between 25 and 50 percent. When combined in large data sets, these findings can guide our efforts to fill the deficits and restore the excess atmospheric carbon to the places where it belongs: forests, soils, and other natural ecosystems. Notwithstanding the urgent need for reductions in human-caused emissions from burning fossil fuels, these deficits can only be filled by conserving remaining old-growth or primary forests, harvesting individual trees responsibly, carefully stewarding managed forests, and restoring forests that have been historically damaged.

Rande and his people and their allies were developing and implementing their vision and plans to regain control of their territory and resources, and to protect, steward, and restore the ecosystems so they could provide economic security for the people once again. We discussed the possibility of placing a Mother Tree experimental site at Hada, to develop Indigenous-led harvesting methodologies and as a place where the youth could monitor the ecosystem responses year after year. As on Haida Gwaii, we knew that any gradient of research sites here and along the Pacific coast would naturally focus on western redcedar, which was the tree of life for the Indigenous people and the species that predominated in the region. We also discussed establishing additional experiments to understand how best to restore the biodiversity, structural heterogeneity, and carbon-rich soils to the industrial plantations.

But we knew we needed to be cautious. Restoring industrial plantations was a new concept, and we'd only established our first trial at Malcolm Knapp Research Forest. Before any concrete actions could

Phillips Arm in Kwiakah territory in various stages of recovery under the stewardship of the Kwiakah Nation. The valley had been thoroughly clearcut by the timber companies over the past century.

be taken for improved management or forest restoration, we needed to spend more time on the land, experiment, adapt from our trials and errors, and learn from the ecosystems.

"It's been a long journey for us," Rande told me. "We've endured so much generational trauma. But we have hope that we can heal from these traumas by reconnecting with the land that has given us life for many years."

I thought of all the years I had spent challenging Western forestry practices and policies, whose main objective was to profit by exploiting the forest at the expense of all else. Academia had been a central part of my identity for decades, but increasingly I was chafing at the limitations of Western science methodologies. I knew some scientists felt I had strayed too far from Western scientific rules and interpretations. But with global change upending our lives and forests dying from fire, drought, and infestations, we needed a transformational shift that embraced holistic stewardship for our collective health and well-being and rejected piecemeal extraction for the benefit of a few. I sensed there was a place for me here with my Indigenous friends, a philosophical and spiritual home. In addition to spending my time convincing critics that forests needed to be respected and understood from a holistic, systems point of view, I could contribute my expertise to Indigenous movements that were already well advanced and were using their own knowledge systems to recover the health of their ecosystems. I could work in service to the first ecologists, the First Peoples, who had originally conceived of and practiced ecosystem-based management.

Healing the land would be no easy task, Rande acknowledged, but "optimism and love would make a hard journey easier." I thought of Mum encouraging me forward. Here, finally, was the good work scientists like me needed to do.

IN THE ARC OF A TURN

It was January 2022 and the snow had been falling heavily all week, with two meters accumulating in the mountains around Nelson over New Year's. The runs were so deep and dense that my niece Kelly Rose, a lifelong skier, had broken her femur on a casual turn. Members of an entire family had broken legs and ribs in an avalanche at White Queen, a precipitous peak just a kilometer from the Whitewater Ski Resort. I would have been skiing, too, if COVID hadn't stung me like a scorpion.

Amanda had just returned from spending Christmas with her family in Prince George. She and a group of friends headed for the irresistible fresh powder at Whitewater. They took the lift up to the summit and skied down a forested glade, where they lost track of one another.

Hannah and I were huddled under blankets on the couch when I heard a knock at my door. I peered out the window to see Rachel and her daughter Magali on the porch.

"We're under quarantine," I called through the glass.

"I need to speak to you." Rachel was insistent. I opened the door a crack.

When Amanda didn't show up, her friends had gone to look for her. They found her near the peak. She had lost a ski and plunged into a tree well, the most hidden and dangerous place in a snowfall. The snow had taken her down, closed over her. Frantically, her friends dug her out, and a nurse who was skiing nearby performed CPR. Amanda still had a faint pulse when the emergency workers took her to the hospital.

They tried to save her, but—

Life, ideas, dreams, all suspended in the arc of a turn, in the gently falling snow. I sat frozen, unable to take in what Rachel was telling me.

Hannah sat next to me. "Mum, are you okay?"

Amanda was just thirty-three. The same age as my brother, Kelly, when he'd been crushed by a tractor after a day of herding cattle at the Mission Ranch, twenty-five years ago. It was early evening in winter then, too. Night had already fallen on the bare-armed aspens.

They tried to save her, but her injuries were too severe.

Death happens as a matter of course in a forest, and we don't wonder about the spirit among the trees left behind. "They're just trees," I recalled a colleague saying to me. Trees drop leaves from the shock of losing a neighbor, then slowly recover and fill in the gap left by the fallen one. Tree mortality, as it's called, fosters biodiversity because a richness of life fills the hole left behind. Gap dynamics create a healthy forest. But losing Amanda was crushing. It was out of sync, out of cycle. It had no conceivable meaning. What could possibly fill the gap Amanda had left behind?

IN A COUPLE OF WEEKS, I tested negative for COVID and my body started to recover from the illness. Yet I found it hard to get up in the mornings, hard to go to bed at night. What was the point? I tried to bury myself in work, writing grant proposals. Trying to convey how we could harness the cycles, the connections between the land and water and living things, everything that felt so abruptly broken with the loss of Amanda.

In the midst of my grief, I continued to get letters from readers thanking me for my book. I tried to respond with gratitude, though I felt numb. Mostly, throughout those dark winter days and nights, I felt hopeless—and angry. At the tree well that had taken Amanda. At climate change, which had caused the phenomenal snowfall with its tempting fresh powder. With my anger came regret. That I had brought her to snowy Nelson, where she wasn't used to skiing around

Amanda, thirty-two, and Eva baked me a chocolate cake with sprinkles for my sixty-first birthday. They also gifted me a new red cruiser vest and stuffed the pockets with pencils, notepads, and treats.

tree wells. That I couldn't do more to help her family in their unfathomable grief. That I'd missed seeing her on my birthday last August.

Amanda and Eva had baked me a chocolate layer cake with sprinkles on top. They had gotten me a new red cruiser vest to replace the special old one that had been stolen from my car, and even bleached it to make it look used and worn-out. Amanda had stuffed the pockets with cans of ale and chocolate bars and my favorite notepads and pencils. She'd gotten me a colorful birthday card with an image of a mother tree with glowing roots. They'd waited for me to arrive, eager to spring the surprise. But I was late, and by the time I got there, Amanda had left. I'd missed her.

I shared my feelings of overwhelming responsibility with Amanda's father, George.

"I get it because I also have those thoughts," he told me. "*If only I did this, if only I did that.* The thing is, ultimately shit happens, and a freak accident that nobody saw coming is just that. Nobody is at fault. We can't live with that fear, otherwise we will miss out on life. Amanda would not stand for that, would she?"

I knew George was right. I tried to lose my grief in the forest, going on long cross-country ski treks, just as I'd done after Kelly died. Day after day, I strapped on my skis and plodded up and down the ski trails in the winter murk, pushing myself to go farther when I was exhausted. Skiing out of Ripple Creek, I sobbed in the steepled firs, the falling snow adding to my tears. The sorrow came in waves.

Then, a month after Amanda's death, we got a letter from the national health service. The doctors had approved Mum's MAID application.

I TRUDGED UPSTAIRS to Mum's bedroom.

"Hi, Suzie," Mum said brightly. Her thin hands grasped the edges of her quilt. I did a little two-step jig, the way Kelly and I used to when we were kids, to make her smile.

Her energy was fading fast after her third bout with *C. diff.* Kelly Rose, having just finished her undergraduate degree in psychology, had moved in to care for Mum around the clock, while Robyn and I

Kelly Rose Heath, twenty-six, and Granny Junebug, eighty-six, in her backyard in Nelson, British Columbia, 2022. We shared many stories, such as the time Mum spent a night in jail after speeding her Camaro through 100 Mile House on her way back from a rodeo in Williams Lake.

visited, cleaned, administered, hovered, wondered, worried. We were all in the same canoe paddling along the shoreline now, the way we used to at Mabel Lake when the cottonwoods were changing color. The current was carrying us along.

"You know, just because you got approved doesn't mean we have to follow through with it. We could wait a year or two. We don't have to do this right away," I told her.

But Mum stayed firm, refusing to back down. She knew what she wanted.

"I want to do this within six months, Suzie."

So, Robyn and Mum and I put our heads together and looked at the calendar to select a date. It felt surreal. We settled on August 15. Mum worried that her last day would eclipse my birthday, which was three days earlier, but I assured her it was okay.

After helping Mum to the bathroom and giving her some lunch, I sat beside her bed. We reminisced about throwing hay at the farm at Edgewood, catching kokanee salmon in the Arrow Lakes, picking lowbush blueberries on the pine flats beside the Kuskanax River, and following the trails to the old Nakusp hot springs. We retold stories of Granny Winnie packing her gun to protect herself against bears. Mum loved reliving these memories and was always ready with more details or a new twist to the tale.

As we talked, my fingers reached for the piece of paper folded in my back pocket. I'd just received an email from some scientists warning they were about to publish an international paper critiquing my research. The three scientists, who had presented their idea at an international conference on mycorrhizas, were coauthors on many of the research papers they were now criticizing. The second author had approached me three decades earlier wanting to learn about forests, as she'd just moved to British Columbia from eastern Canada, where she'd worked on mine reclamation. I had welcomed her and her graduate student—the lead author on the critique—to my home forests and helped with their funding and research. Jean and I had located their experimental sites, taught them about the forests, and showed them how to plant seedlings in the forest floor, not the skid roads, so the trees would survive.

Now they were questioning my experimental techniques, my results, my interpretations, which were about the same type of connections that had enabled their experimental seedlings to survive. The papers they were criticizing had been published in international journals for over a decade, in some cases almost twenty-five years earlier, each peer-reviewed by three to six reviewers. I couldn't understand why they were attacking these papers so many years after they'd been published.

Mum was in her final months and my daily problems were insignificant compared to the end of life. But I was her daughter and I needed her advice. She was still my mum and she listened as I shared my qualms.

But I could tell she was getting tired. I changed the subject and we reminisced about climbing the willow in our backyard when I was four, reaching as high as I could to touch the sky. When at five I'd gone up the enormous cottonwood tree to see how far I could go before I fell out onto the roots. And when at six, I'd learned about roots and mushrooms while Grampa Henry dug our dog Jiggs out of the outhouse.

"Remember the night Robyn hurt her toe?" Mum said.

I smiled. The memory was still vivid.

It was summer, and our family was spending our annual two-week holiday at Uncle Jack's logger's houseboat at Cottonwood Bay on Mabel Lake. Robyn was eight, I was six, and our little brother Kelly was four. We kids spent endless hours playing on the dock, jumping into the lake, and waterskiing behind Uncle Jack's boat. Grampa Henry had lent Dad one of his handmade riverboats—not his prized *Saint Laurence*, but a close second. It was a perfectly crafted wooden boat, waterproofed with pine resin and stained dark brown.

One night after dinner, Robyn was helping Mum clean up when she accidentally dropped a chair on her big toe. It hurt like hell. My parents worried all evening, exchanging whispered conversations as they watched Robyn's toe start to swell and a dark purple bruise form behind her toenail.

The houseboat was a good two-hour boat ride away from the Shuswap River mouth where Grampa Henry and Granny Martha lived. Maybe Robyn would be fine. But by midnight, my sister was in pain. The blood was causing a lot of pressure under her toenail. My parents fretted, uncertain. Should they wait out the night, or risk the trip across the lake to get Robyn to a doctor? Crossing the lake in the pitch-black night was risky. What if a storm blew up, or it started to rain, or the engine quit, or we capsized? No one would ever know what had happened to us.

Finally, they made a decision. They bundled us into our life jackets and blankets, and we all got in the boat and putt-putted out onto the dark lake. Our only light came from the moon and stars. My parents were quiet with fear. It would take a couple of hours to get to Grampa's dock, and then a forty-kilometer drive along a rough dirt road in our old Pontiac from the river mouth to the hospital in Enderby.

The air was still as we approached the rock wall where the Splatsin people had painted pictographs of their cosmology long ago. Suddenly, the outboard motor cut out. Mum gasped. Dad struggled to restart the motor while we sat in the dark, drifting.

"I think the cotter pin has broken," he told Mum.

Dad couldn't swim, but he didn't say a word, his mind trained to remain calm by generations of *scieurs de bois*, our ancestors who'd emigrated from France to Quebec, then made their way across Canada to practice their trade horse-logging in the inland rainforests of British Columbia. We sat in silence, the generations of trees in our blood. Maybe we could get to the river mouth by paddle. Robyn was in increasing pain, Mum on the verge of panic. Suddenly Dad reached under the wooden seat and grabbed something, then opened his palm to show Mum. *A new cotter pin.* Grampa Henry had taped a spare under the seat. Dad held it up with a grin, then replaced the broken one and grabbed the pull cord. We putt-putt-putted into the night.

We reached the river mouth, where the river ran fast and our cousin Katie Mae had once slipped and fallen in. In the seconds before the current swept her under the wharf, her mum, Auntie Yvonne, grabbed her ponytail and hauled her out in the nick of time. The beam from a lone flashlight sliced the darkness—Grampa Henry must have heard us coming. He helped us out of the boat and we bundled into the car and drove through the night to the Enderby hospital. A doctor met us there.

"He used a Bunsen burner to heat up a paper clip, and he burned a hole in Robyn's toenail to release the pressure," Mum remembered.

"It was dawn by the time we got back to the houseboat," I said.

Her eyes were closing, so I pulled up her covers and shut the curtains, got ready to leave.

"Suzie," she said.

I turned around.

"You come from the forest. The trees are in your blood. No one can destroy the knowledge and understanding inside of you."

Then she fell asleep.

IN THE MONTHS FOLLOWING Amanda's death, the many people who had loved her found ways to honor her memory. Her parents, brother, and sister-in-law, her community, her sports family, her academic family. Friends, teammates, lab mates, colleagues. Teachers and coaches, the young athletes she'd coached. A community foundation in Prince George created two annual memorial awards in her name. The university started collecting funds for a scholarship in her name that would support female athletes in academia. One of her former coaches called her the greatest player in the history of Canadian women's baseball. She was inducted into the British Columbia Sports Hall of Fame, the highest sporting honor in British Columbia. The women's national baseball team retired her number—19—the first time they'd ever done so. Amanda was irreplaceable.

Meanwhile, Mum had started preparing Robyn and me and all of her grandchildren for her departure. She reassured us that she was simply following the natural cycle of life. I didn't know if I could bear another loss so soon, but this was Mum's journey, I reminded myself. She was moving along her own path, on her own terms.

"I will be going to infinity," she told me.

It eased my mind to imagine Mum in an infinite cycle of passing from one life to the next, elder to child to adult and back, with some magical energy transformation in between. A sort of reincarnation. I understood the forest in this way too. Not linear, but cyclical. When elder trees die and decay, they become part of the food web for new generations of seedlings that grow and mature into elder trees—a cycle of life that continues to infinity. Thinking of Mum's life in this way reminded me of the adaptive cycle of complex systems that scientists had visualized using the infinity symbol. All living systems, including humans, revolve through stages of growth, conservation, reorganization, release, and back to growth, ad infinitum. Mum was at the release stage, preparing to pass her final words of wisdom to the young.

One sunny morning, Eva dropped by the house with Tia. We had coffee on my back porch and Tia kept a watchful eye on the resident squirrel in the maple tree. Through tears and laughter, we shared our stories about Amanda.

I grieved for the bright future Amanda had been robbed of, that we had all been robbed of by her loss. She had died too young, with so much left unfinished. We would never know what scientific questions she might have asked, what contributions to stewarding our forests she might have made. Even so, she had left a remarkable legacy. She had discovered that Douglas fir recognized and favored selection traits of nearby kin seedlings over strangers, and that the species composition of the whole forest influenced these interactions. Her questions had grown from her wisdom that it takes a community to raise the next generations, and that relationships among individuals helped to build community. That was the philosophy that had made her such a beloved team captain and cherished lab mate.

I thought back to the difficult days when I'd been diagnosed with breast cancer in 2012. I'd started chemo while Amanda was doing her master's, and I'd had to take a leave of absence from the university and step away from the lab. Amanda and the other lab members had come together to run it. Amanda had served as captain when needed, while also sharing the leadership role with others. The students took turns leading, holding things together, supporting one another, and keeping the lab running, shifting roles just as trees do in a forest when one starts to succumb from disease. Strength and weakness arise at different times in the life cycle of a forest, just as in the life of an individual. Community is what makes a forest—and a person—whole, unique, and resilient. Amanda had understood that.

Eva and I tipped our mugs to Amanda. As we sipped our coffee and stroked Tia's silky coat, we reflected on the beautiful cycles of life and death, knowing how difficult they were to embrace when they had touched us so directly.

14

TREES IN OUR BLOOD

That summer, Robyn helped Mum put her affairs in order and arranged for friends and family to visit and say their goodbyes. This was Mum's last month. Tears were mixed with laughter over stories of Mum in her lime-green VW van, the back hatch spray-painted with the words "Abominable Slow Van," winding her way over the mountain passes to go windsurfing with friends. Or roaring up to the Whitewater ski hill, hell-bent on finding rock-star parking so she could be first on the chairlift for fresh powder on the Huckleberry run.

In a quiet moment when the preparations were in hand, I made a short trip to the ancestral Kwakwaka'wakw village site of Hiladi, or "The place to make things right," located in Johnstone Strait on Vancouver Island. There the Mother Tree team would join Rande and members of the Kwakwaka'wakw Nation and friends for the annual gathering of the Awi'nakola Foundation, an eclectic alliance of Ma'amtagila people, artists, scientists, and allies dedicated to healing forests and reviving cultural sites.[1]

My students and I were there to document the effects of clear-cutting on the trees, plants, and soils. We hoped that this baseline data—recording what had been taken and what remained—would be helpful in returning the land to the Ma'amtagila people and restoring their forests with species important to them that could adapt to the changing climate. The data we gathered could support the people's efforts to maintain their own laws, language, and lifeways on their own terms.

AFTER TWO HOURS of bone-jarring travel on rutted logging roads down to Hiladi, we parked our trucks where the trees were thickest and the salmonberry tallest. We walked down a narrow path to a clearing by the sea, the site of a once-bustling village whose people had been forced out after European settlers brought smallpox and other diseases in the nineteenth century. A fire was being tended by Ma'amtagila matriarch Tsastilqualus Ambers Umbas, whose first name translates to "Where people gather around the fire for warmth." Tsas had been living on the land for over a year, her presence an enduring symbol of her people's sovereignty and their yearning to reconnect to their land and villages.

The immense, monumental trees that had once stood there and the creatures who depended on them were now gone. But with one person at Hiladi, Tsas believed, more would come. Members of the Ma'amtagila Nation who lived in Alert Bay, Campbell River, Victoria, and even farther away would return to their traditional territory to tend to the salmon streams and care for the root gardens. Then the shrubs, herbs, mosses, and trees, the ermine, bears, and deer, and the salmon, fishers, and sea otters would also return to their ancestral homes.

"We have been gone, but now we are back," Tsas said. "This is my home and I will never leave again."

As we sat around the fire, she told us about the matriarchs, youth, and supporters who were joining her to oppose the commercial fish farms and industrial logging in their territory. A crew of volunteers at the University of Victoria had constructed a tiny house and then moved it to Hiladi to serve as a home base, where it was dubbed the "Little Big House." Tsas described plans to add a bunkhouse and an outdoor kitchen. She had to raise her voice above the backup alarms coming from the logging trucks and loaders working just a few hundred meters across the confluence of the Adam and Eve Rivers. The last of the great trees were being felled, limbed, and loaded, stacked up like dollar bills.

While Tsas was confident this renewal would come, I knew there was much work ahead to recover the vitality of these ecosystems. I was astounded by the extent of the clearcutting. Other than a few patches, the entire forest had been taken. The trees that had been densely planted by the logging companies in place of the cathedral forests were thin, gray, and scaly. The floodplain, once rich with plants in the understory of giant cedars and cottonwoods, was devoid of biodiversity. Extensive spraying of glyphosate had killed any native herb or shrub considered a threat to future profits.[2] The devil's club leaves were yellow and the salmonberry lacked fruit. Huckleberries and strawberries hid among the shadows, while the elderberries and thimbleberries were confined to the roadsides. There was no room for blueberries, currants, or cloudberries. I couldn't see any golden chanterelles, king boletes, or true hedgehogs, the mushroom species associated with old trees.

"It pains me to know that all of the traditional medicine is being destroyed," Rande said, pointing to some devil's club that had escaped the spraying. "This practice is banned in Quebec, yet we're still spraying it here in B.C."

My years of research in the inland forests had shown that glyphosate did not always improve the growth of conifer plantations as expected. Even worse, this long-standing practice was often associated with loss of the prized planted trees to disease and insect infestations. The herbicide spraying also harmed riparian species in the streams and wetlands and increased the risk of fire, infestations, and flooding in the watersheds. The erasure of diverse habitat not only killed the native plants, but also reduced populations of mammals, birds, and fish.

Hundreds of species had once interacted and collaborated here to create a healthy, productive ecosystem. The roles formerly occupied by all of these species were now empty, the niches flattened and compacted down to a single conifer species, planted decades ago after the forest had been clearcut. Looking over all of the empty niches, I began to envision how we could help. The data from the Mother Tree Project had shown us that the mother trees were the energy centers of the forest and were critically important for regeneration. This knowledge could help us to restore this forest, by leaving some of the bigger trees standing while selectively thinning the densely planted understory to allow light to penetrate to the niche spaces so they could fill back in.

We understood so little of this new craft, the restoration of industrial plantations, and we would be learning as we went. My first instinct had been to thin out the many hemlock recruits to release the big trees. But my master's student Kiku Dhanwant would soon make the discovery that hemlock saplings crowding the dense understory of second-growth forests on Haida Gwaii played crucial roles as emerging snags. She observed that red-breasted sapsuckers fed on the hemlock sapwells, introducing decay fungi and slowly killing the younger trees, and that these new snags were contributing to the heterogeneity of the stands and, in time, creating niches for old-growth species. Kiku's observations reminded me that effective restoration practices would require careful experimentation, monitoring, and adaptation as we learned more about the ecosystems.

Rande, Teresa, the crew, and I discussed techniques we might use to repair the riparian communities along the salmon streams and make room for the food and medicinal plants. Thinning the weakest trees from the dense stands while favoring the largest cedars, hemlocks, firs, and spruces should allow the rich understory to return and prime the seed banks, bud banks, and soil food web. This restoration would not be simple, as it depended on the unique qualities of the place and the goals of the nation. It would depend on the essential lay of the land, the innate patterns of the water flows, and the constituent fluxes of the nutrients. The work would take into account the properties of the local soils, the intrinsic requirements of the plant communities, and the fundamental needs of the area's wildlife. We would design the thinning to encourage the community of species that fit the ecosystem, culture, and changing climate.

Our research suggested that we could thin the wetter lowland sites more heavily than the drier upland forests because of their great productivity. We would carefully remove the smallest trees by hand-falling, leaving the largest trees in densities and patterns that favor rapid growth and provide varied structure and niches for returning species. This would encourage advancement toward conditions resembling an old-growth forest, and the roots and crowns of the remnant trees would nourish the gaps opening up. The understory plants would fill the niches, capture the resources, and connect with their neighbors. In turn, they would produce berries, nuts, seeds, and pollen, and nourish the soil. Their ability to convert the full spectrum of sunlight would produce photosynthate for the whole heterotrophic food chain and help restore organic matter to the soils. Wood from the thinnings could be burned for heat or used to make homes and smokehouses. Where this was not possible, it could be turned into charcoal, or biochar, and returned to the soil for long-term carbon storage.

In the uplands, we would thin the stands more lightly to encourage huckleberries, tall Oregon grape, and salal, prized as traditional foods, and to provide protective places for medicinal plants like devil's club, yarrow, and kinnikinnik. We would need the overstory cover at the headwaters to capture the snowfall, filter the rainfall, and stabilize the soil, helping restore the hydrologic integrity of the watersheds. With the living soil better able to absorb the water, the streams would return to healthy flow cycles and lower the risk of flooding or drought.

Reviving the ancient fishing technologies, working with the streambeds, weirs, and traps, and tending the clam beds could help nourish the gardens near the village, where the silverweed, springbank clover, and chocolate-petaled riceroot once flourished. The seeds, buds, and roots of the root plants waiting in the soil would sprout and prosper under the care of the Ma'amtagila people. My team could bring in instruments to measure the recovery of the webs, communities, and cycles. Western governments and scientists responded primarily to hard data, and it would help Rande build his case for rights and title in the courts.

As we were leaving Hiladi, Tsas turned to us.

"I'm where I am supposed to be. I don't have to ask permission from anyone to return to the land our people are from. I belong here. My job isn't done."

Back home in Nelson, Mum was spending time with her granddaughters. Hannah had just learned she'd be helping the Simpcw Nation reforest parts of their territory damaged by the building of the Trans Mountain pipeline. She was puzzling over the selection of shrub species that served as important food plants and also helped to improve the soil. Nava would be doing her graduate studies with the Nlaka'pamux ecologist Jennifer Grenz, who directed the Indigenous Ecology Lab at UBC. She would be mapping plant species in the Cowichan estuary on southern Vancouver Island as part of a project to revitalize the intertidal food system of the Quw'utsun people. Mum wanted to hear all about their work.

I headed for Elephant Mountain across the lake and walked up my favorite trail, a salve for what was to come. I sat among the Douglas firs and ponderosa pines, their roots entwined and cambium conveying notes of sweet vanilla. A squirrel carried a cone to his cache, hiding among the saplings nestled under the crowns of the old trees while biting off scales to get at the seeds. A dried mushroom hung from a branch, the spores sprinkling over the seeds in the midden. The cycle was all here before me.

My research on common mycorrhizal networks was part of a transformational shift in how we understand and manage forests. I endeavored to show that, far from just a collection of individual trees competing for resources, a forest is a complex adaptive system aligned with the Earth's natural connections, relationships, and cycles. That interdependence is an essential feature of thriving living systems. The goal is to transition from a focus on timber extraction toward a model that prioritizes ecosystem health, and thus stave off the worst losses of the climate crisis. This means protecting primary and old-growth forests and practicing proforestation—allowing intact forests to live to old age and reach their ecological potential. It means better management of secondary forests set aside for wood products and restoring damaged ecosystems for the many amenities they provide.

Whenever there is a paradigm shift, however, there is resistance. If I'd been unprepared for the positive popular reception to *Finding the Mother Tree,* I was even less prepared for the backlash from the small but vocal group of scientists. They argued that my lab's research had discounted the role of alternative belowground pathways to common mycorrhizal networks for resource transfer. The three scientists claimed that our research ignored the important role of competition in forest dynamics. They implied that the carbon transfer we'd measured between trees was insignificant because the isotope did not enter the shoots of the recipient seedlings. And they maintained that none of these findings had been associated with the regeneration of seedlings in the forest.

I was mystified by these claims. My articles—which were all peer-reviewed and published in international scientific journals, as was the common practice—consistently referred to multiple belowground transfer pathways, including through common mycorrhizal networks as well as the soil, roots, or disconnected fungal networks. Even more puzzling, none of our lab's articles had negated the process of competition in forests. Instead, we showed that numerous types of species interactions—including competition and collaboration—took place simultaneously in the forest. Our studies repeatedly demonstrated that significant amounts of carbon were transferred into both the shoots and the roots of the recipient seedlings.[3] And many studies besides ours had shown that common mycorrhizal networks were essential to the establishment of seedlings and mycoheterotrophic plants— plants that form a symbiotic relationship with fungi to get most or all of their nutrients.

But perhaps worst of all, in their view, I had committed the cardinal scientific sin of anthropomorphizing trees by using the metaphor of the mother tree to represent the role of the oldest, biggest trees in the regenerative capacity of the forest. The mere utterance of the word "anthropomorphic" in Western science is meant to be dismissive. It suggests the scientist who makes this blasphemous mistake is not an objective observer, but rather impure, intuitive, and subjective, perhaps lacking integrity. Perhaps too emotional or spiritual. Too feminine.

The three critics were so incensed by this that they discounted Amanda's discovery of kin recognition in trees—their rationale being that the precise belowground mechanism by which trees recognize their genetic relatives remained uncertain. Yet our research had demonstrated that this process had significant effects on the adaptive traits of Douglas fir seedlings and could provide insight into why natural regeneration can often outperform planted forests under climatic stress.

I was well aware that critique was a crucial part of the scientific process. My area of study had been a subject of lively peer review. It would, and should, continue to be scrutinized. Still, I wondered how much proof these scientists would need before they understood that the forest was a connected place. I remembered Teresa saying, *We've known these connections for thousands of years.* Of Rande telling the kids, *We are interdependent with the trees.* A forest could only be understood as a continuous, integrated living system. My own life and work had shown me the truth of this.

Decades of experience had taught me that we needed to shift our worldview from commodifying to caring for Earth's ecosystems; to respectfully steward instead of exploit our forests; to address not just our material needs of the moment but our responsibilities to the greater good and to the future. The question we scientists should be tackling was how to better care for our forests to ensure a livable planet for coming generations rather than pointing fingers at one another. The people choking on smoke from fires in Portugal, Greece, and California or watching beetle infestations and drought kill woodlands in Europe and North America needed scientists to better predict and mitigate such emergent phenomena to protect their lives and livelihoods and those of their children and grandchildren.

There in my home forest, among the pines and firs on Elephant Mountain, I had an epiphany.

As scientists, we are taught to be reductionists, trained to pick apart and study one process at a time, even though we know these processes don't happen in isolation. Science is a powerful tool to learn about the pieces of systems, but that knowledge is of little value without the broader context of ecosystems, people, and cultures. Reductionist science, which emerged during the Age of Enlightenment in the seventeenth and eighteenth centuries, places a premium on ideals of competition and hierarchy. It's good at studying objects, but struggles with more stereotypically feminine or Indigenous concepts of relationships, collaboration, and networking.

I had become interested in common mycorrhizal networks because I saw them as a system of relationships and connections that underscored forest health and resilience. The interdependencies these linkages represented made intuitive sense to me. I knew the forest from a holistic perspective, understanding the whole first and then seeing the parts. I had inherited this way of seeing the forest from my ancestors. I understood the complexity of its relationships, and how these interactions created vitality, not just in individual trees or fungi or carbon, but in the whole system. I listened to the teachings of the common mycorrhizal networks using the Western science methodologies, and then I translated the learnings into language that nonscientists could understand.

I was sure my critics loved the forest, too. Perhaps they walked and communed among the trees just as I did. But their critiques showed they saw the forest first as a series of parts, and they were trying to add up the parts to a living system. They claimed there was no proof that big, old trees were the energetic centers of the forest, even though that fact was obvious to anyone walking through an old-growth forest. This compartmentalized approach to studying complex natural ecosystems was why the critics had concluded there was "insufficient evidence" to make changes to forestry practices, even as widespread clearcutting degraded the forests and wildfires blackened the countryside. They simply could not find the whole by adding up the parts.

MUM'S LAST DAY was a beautiful summer day, and I felt sick to my stomach. I hadn't been sleeping well, and I was exhausted. I kept thinking she would change her mind at the last minute, or that we'd figure out some way to keep her with us. But she was set in her decision.

We had organized a ceremony so we could say goodbye and honor her in her last moments. Kelly Rose put together a playlist of Mum's favorite Willy Nelson songs, and Robyn figured out where we were all going to sit and who would read a poem or tell a story. Kelly Rose bathed Mum and helped her put on the outfit she'd picked out—her skirt with tiny blue forget-me-nots and her white silk blouse. She settled into her rocking chair on her front porch, and we all gathered around and said our last words to her. I read her favorite poem by Joyce Kilmer. *I think that I shall never see / a poem as lovely as a tree.* I felt like I was in a dream. Could this really be happening?

After lunch, I took her upstairs to the bathroom. She climbed the stairs slowly, leaning heavily on my arm, stopping at each step to catch her breath.

"Suzie, I just can't do this anymore," she said. "I'm ready to go."

I remembered how fatigued I'd felt in the days following chemo, so worn-out I could hardly walk another step. I'd been lucky and recovered, but Mum wouldn't be recovering. She wouldn't be getting any better. Things could only move in one direction now.

When we came back downstairs, the doctor and his assistant had arrived.

"Hi, June. Do you know why we're here today?" the doctor asked gently.

"Nope," Mum said. The doctor's eyes widened. "Oh, I'm just kidding. I know why you're here."

Everyone laughed. That was Mum, up to her usual tricks.

Mum sat back down in her rocking chair and we hovered inches away. Nava leaned into me and Kelly Rose. Robyn held Mum's hand. The doctor asked several more questions to make sure Mum knew exactly what to expect. He checked with her one last time. "Are you sure this is what you want, June?"

"I'm sure."

The doctor carefully placed the IV line in her arm, but as he did, a little spot of blood beaded up and Robyn and I exchanged a look, because we knew Mum was terrified of pain and we tried to comfort her.

But Mum just said, "Oh, you girls. I'm fine."

The doctor gave her the first of three medications, a tranquilizer to relax her.

"Is it over?" she said. "Well, that was easy."

"Oh, no, we're just getting started," he said.

I could see that Mum was getting drowsy. As he administered the second medication, her eyes fluttered shut. She started talking in her sleep, so quietly that I had to lean forward to hear her murmur her last words.

"I can't climb the apple tree anymore. I can't lead the horses up the mountain anymore."

I thought of Tsas saying, *My job isn't done.* Mum had done her part in caring for her people. She had tended her garden, raised her children, and cared for her grandchildren. She could no longer do the things that gave her life purpose and joy. Her job was done.

The doctor gave her the third medication, and within minutes, Mum was gone. Her spirit free, she was off to infinity.

THREE DAYS AFTER MUM DIED, Nava and I hiked to the Silver Spray Cabin in the Selkirk Mountains near Nelson. Exhausted from the emotional strain and sleepless nights of the previous months, I needed some respite from the sadness, to breathe mountain air again. With the losses of Amanda and Mum within the space of a year, I had been in survival mode. I'd been pushing my grief away, promising to tend to it later. But later never came. Tears would come at the worst possible times.

The day we left, I learned that the article critiquing my work had been submitted to a scientific journal. Despite my attempts to push the criticisms to the back of my mind, the consequences had been steadily mounting. Two letters had been sent to my university claiming I lacked "scholarly integrity," an accusation surely intended to have me fired. Someone called on the CBC, BBC, NPR, and National Geographic to withdraw their profiling of my work. Warning messages arrived in my email inbox that my research "crossed a line" and to "step back on the limb you've climbed out on." A colleague turned down a request to voice support for Amanda's research findings on kin selection. The media, sensing clickbait, leapt upon the controversy and branded it "a bitter fight among scientists." But I did not want to fight.

I slung on my backpack full of hastily chosen food and wine, and Nava and I headed along noisy Woodbury Creek before veering sharply upward through a recent burn. I knew the trail from the year before, when the summer was mild and the huckleberries abundant and sweet. The switchbacks were steep, climbing about 1,000 meters elevation over 7 kilometers (3,300 feet in 4.5 miles). We had three hours before dark. I had decades of backpacking under my belt, though none recently, of course. *You've got this, you can make it,* I told myself. Nava, strong as a deer, loped up the mountain in front of me. I flashed back to being in my twenties, when Jean and I had trekked through the forests like mountain goats, scrambling up and down boulders and scree slopes, with little thought to the risks of falling or injuring ourselves.

As Nava and I climbed up the switchbacks, the heat drew streams of sweat from my tired body. The summer of 2022 once again had broken maximum temperature and drought records, spawning lightning storms that ignited fires through the mountain ranges. The soil was so dry that dust kicked up and mingled with the smoke, and I gasped in the thick sultry mixture. My pack was heavier than I ever remembered a pack being, and I had to stop at each switchback to rest. Nava slowed her pace to match mine.

Halfway in, I was stopping every twenty meters to catch my breath. This must be how Mum had felt these past few months, I thought, her pace having slowed day by day. Nava took some of my gear to lighten my pack and helped me refocus by pointing out the expanse of white pearly everlasting and magenta fireweed bursting from the burned, blackened soil. Among them were new lodgepole pine and subalpine fir seedlings, their seed dispersed from old trees protected from the fire in a nearby hollow.

"Look how resilient the forest is, even after that wildfire," Nava said.

I felt relief that the forests were coming back after so much stress. They had evolved for healing and recovery, but climate change was creating new conditions unlike anything we had seen before. In spite of this, there remained good soil, well-adapted seed, and nearby trees to bootstrap the ecosystem back to health. *If the forest can bounce back, so can I.*

Still, I couldn't pick up the pace, and dusk was coming. Between the smoke and thoughts of Mum and Amanda, I was struggling to keep the tears back. Each step felt like climbing a sheer cliff with a sack of rocks on my back. Nava started to look worried. We still had about an hour to go before we got to the cabin. Normally I was the strongest hiker, so she knew I was not myself. We stopped at a grassy patch to put on our headlamps and I glanced around for tent spots, but the slope was too steep. A marmot popped out of a hole and whistled, seemingly to urge us on. We trudged forward, one slow step at a time, Nava checking me at every switchback. Just as darkness was falling, high in the alpine, we made it to the hut.

We ate dinner overlooking the valley. Through the dusky sunset,

Fungus in a continuing cycle of life and death and new life, regenerating year after year, in balance with the cycles of the forest.

I admired the complexity of the forest interwoven with the dramatic contours of the mountain ranges, reminding me that each patch of trees, meadow, and avalanche track was finely attuned to its place. Each creature lived its entire life cycle in these forests, adapting, evolving, and responding to the changes in weather, and this enabled regeneration, longevity, and resilience. These forests were designed to heal. Mum had been born, lived, and died in these mountains. She too was part of the fabric of resilience.

The boxed wine I'd brought tasted terrible, but we made a toast anyway.

"To Granny," Nava said, and we tapped our cups.

"She made it to the other side," I said.

I gave Nava a hug and we sat for a long time looking at the stars before falling asleep.

The next day was clear, and with no packs to carry, we scampered above the cabin onto a ridge. My vitality had returned, even on the steep slopes. We climbed up Mount McQuarrie, then down a small pass, then up another unnamed ridge, and looked over the valley at the slide where I'd seen a wolverine the previous year. The purple flowered penstemons nestled among the rocks and the tough little subalpine fir leaned over, stunted from the wind, adapted to live with strength.

My thoughts kept returning to the paper my critics had written. I knew I was far from the first scientist who had challenged the old guard. In some ways, it was a sign of progress. Scientists had sneered half a century ago when Jane Goodall gave names—*David Greybeard, Flint, Frodo*—instead of numbers to the chimps she was studying in Gombe. They were scandalized when she drew parallels between chimp and human behavior, and when she observed that each chimpanzee had its own distinct personality.[4] Scientists had been hostile to James Lovelock's Gaia hypothesis that the Earth is a synergistic, self-regulating system, much like the human body. Lovelock was accused of being anti-Darwinist, a neo-pagan, too metaphorical. Lovelock's colleague, the evolutionary biologist Lynn Margulis, was disparaged for her theory of endosymbiosis to explain the origin of eukaryotic cells, which later gained widespread acceptance. Margulis also helped reframe Gaia as a concept, not an organism, but "an emergent property of interaction among organisms." "Life on Earth is more like a verb. It repairs, maintains, re-creates, and outdoes itself," she famously said. I didn't feel like I belonged in such illustrious company, but my spirits were buoyed at the thought of these scientists who had persevered in the face of criticism and scorn.

Nava and I picked our way down from the ridge through a steep, snowy notch. The slope was over the angle of repose and a slip could result in a long, dangerous slide. I talked her through each step, helping her find her footing down each reach. It took an hour, but we finally made it to a safe spot to rest. Things were back to the way they should be, I thought—mother helping daughter. But these roles were fluid, and now Nava was mature enough to help me, too. I told Nava about Mum's anxiety walking over boulders many years ago in the Stein Valley, and how I had helped her calm down and make it over the talus slope.

We stopped for lunch by an alpine lake. The sun was warm and we sat silently, the sound of the waves lapping against the shoreline soft and rhythmic. I watched a blue dragonfly flit about in the willows and thought of Amanda, her laughter and her kindness. I thought of the *nawalakw*—the supernatural forest—and of Rande telling us about the teachings of the trees, how we all hold one another up.

Mum had been here for a moment. She would always be here with us, in the wind and the trees, teaching us to be open and flexible and to give it our all. Amanda would be with us, too, holding us up, inspiring us to give others our support and encouragement and make them shine.

I would carry them in my heart and keep them close.

BACK TO THE GARDEN

The crowns of millions of trees reached up toward the sun, reflecting every possible hue and luminescence of green, absorbing the entire spectrum of photosynthetic light. They were cycling billion-year-old carbon, sequestering 140 billion metric tons of the ancient element in above- and belowground pools, while stabilizing Earth's climate. In the process, they spiraled out oxygen, which I gulped in excitement.

I was in the ancestral territory of the Achuar people, near the border between Ecuador and Peru, more than 6,500 kilometers (4,000 miles) away from my home in Nelson, British Columbia. Three-quarters of the Amazon basin's rainfall—2 to 3 meters a year—cycled back and forth between the forest and the troposphere, the lowest layer of Earth's atmosphere.[1] The rest was in the muddy Amazon River, whose path slouched through the landscape to the Atlantic, releasing a staggering 112 billion liters (29.5 billion gallons) of water a minute—so much fresh water that 150 kilometers (93.2 miles) out to sea from the river mouth a person could drink from the ocean.[2] The river gave life to scarlet macaws, pink dolphins, jaguars, and humans. It fed mammal, bird, amphibian, reptile, and fish species numbering in the thousands, plus two-and-a-half-million insect species and tens of thousands of plant species.

My imagination had not prepared me for such abundance. I exhaled in reverence. The Amazon rainforest had lifted me from my pool of loss into the delight of the world again.

It was the first anniversary of Mum's death, and I'd been feeling tired and discouraged. If anything, the criticisms of my research had grown more insistent over the past year, and it was taking a toll. I'd started avoiding my workplace and dodging my colleagues. On Zoom calls, I kept my camera off and rarely spoke. When I was on campus, I stayed in my office and avoided walking down the hall. I worried about Eva and my other students, who were watching this campaign in dismay. I was concerned about my daughters, who were just beginning their graduate studies and professional work in forest science.

Hannah was working deep in the forests at the northern reach of Kinbasket Lake, a reservoir on the Columbia River in southeast British Columbia, and I panicked one night when she hadn't called to check in with me as planned. Later I learned that her crew was stranded on the shore by a windstorm, the boat unable to navigate the choppy waters to retrieve them. They'd had to hike through the night to a place where the boat could safely motor in, then ride back with one-meter-high waves spilling over the bow. When she finally called, the sound of her voice on the other end was the sweetest thing I'd ever heard.

I'm not built for this, I thought. *I didn't sign up for this.* Yet I could not just walk away. I was no longer just an anonymous scientist studying my own home forest. Through my research and scientific publications, my TED Talks, my book, and the launch of the Mother Tree Project, I had become a public figure. I had responsibilities, obligations. People were looking to me, relying on me to help protect the trees. I could not just disappear.

I was in this demoralized state of mind when I received an invitation to take a cultural trip to the Amazon with an organization called the Pachamama Alliance. While the Amazon rainforest was far away from Canada's inland temperate forests, I was eager to see the plant and animal communities that thrived in the tropical forest and to learn from the Indigenous tribes with whom the Pachamama Alliance collaborated.

WHEN OUR PLANE touched down on the red earth landing strip a hundred kilometers (sixty miles) into the jungle, a group of Achuar men, women, and children stood by a wooden thatched hut to greet our small group of travelers. The children, shy and curious, playfully escorted us off the runway. The men led us down a wide trail to the river, and the women watched over the process, their babies tied to their sides.

Ramaro, our Achuar guide, welcomed us and seated us in a narrow wooden riverboat called a *voadeira*. We motored up a muddy tributary to our lodge in the Sacred Headwaters of the Upper Amazon. A latticework of more than a thousand smaller rivers served as the highways of the rainforest, for getting from one village to another, for transportation and trade, for fishing and hunting. Turtles lined the tree limbs overhanging the shoreline, and a capybara, a rodent at least as big as me, glanced sleepily at us from a sandbar. A pink river dolphin rose in a smooth arc beside the boat. The *voadeira* reminded me of the riverboats that Grampa Henry had made to ply the waterways of Mabel Lake back home.

As we glided up to a small wooden boat launch, Ramaro hopped out, reaching in to help the rest of us clamber onto the landing. With my duffel bag slung over my shoulder, I stepped onto the path and looked up in awe at the enormous mahogany, strangler fig, and balsa trees towering overhead. With my head tilted skyward, I absorbed the intricate patterns of the crowns, branches, and leaves. I was mesmerized by the prodigious size and structural complexity of the overstory trees and the dazzling diversity of the understory plants, which I knew were fundamental to the functioning of a healthy forest ecosystem.

Eventually, I lowered my gaze so I could more fully grasp the fractal geometry of this forest. Every angle of life had exquisitely evolved to capture the full spectrum of light, water, and nutrients. The space between the understory and the overstory—what ecologists called the subcanopy—was packed, every niche filled with saplings, ephiphytes, bromeliads, lianas, and vines. The subcanopy, I would learn, was where most Amazonian tree diversity existed. Differences in stature of these plants, and in their resource requirements and tolerances, allowed them to permeate the entire height gradient. The plants and shrubs—heliconias, cacaos, orchids—were even more diverse. Variation in shade tolerance allowed these diminutive species to partition the horizontal light gradient at the forest floor.

While my eyes were saturated, my ears were filled with music. Millions of insects whirred, clicked, and buzzed; toucans and macaws called from leafy crowns; howler, capuchin, and spider monkeys barked, hooted, and screeched from the canopies. Vibrantly colored beetles as large as my hand chirped and trilled, and I wished I knew their names. Tarantulas hiding under massive leaves and populations of spiders covering thin threads purred, hissed, and rasped and had me jumping back. Leafcutter ants trooped in long lines from trees to deep caverns, carrying pieces of cut foliage that rattled like shopping bags, and I watched them as long as I could. They were cycling the photosynthetic energy from the trees deeply in the earth, then carrying back bits of living soil to prime the cycle.

Mesmerized by the constellation of diversity, it took me an hour to walk 50 meters. I finally stepped inside my palm-thatched hut, which I was sharing with a woman from my group named Peggy, and a cockroach scurried across my foot. I screamed, she screamed, and the cockroach probably screamed, all of us shocked to find ourselves caught inside this little room together. Peggy scooped up the three-inch insect and transferred it outside, instantly cementing our friendship.

By the next day, the patterns of the jungle had taken over my mind. The architectures of the tree crowns, the motifs of different-ordered branches, and the adventuresome roots all fit together in geometric formations to absorb every available resource. The spacing of the largest trees, frequency of disturbance gaps, and regularity of tree mortality were interwoven together in complex designs driven by disturbance, dispersal, and density. The images became etched in my mind's eye as I stared at them. *Why this pattern, this diversity?* When I closed my eyes, the images were still there. When I fell asleep at night, they filled my dreams.

THERE IN THE JUNGLE, Mum and Amanda stayed in my thoughts. I felt their presence in the beauty that surrounded me, in the motifs of branches, roots, and fungal networks. On my walks in the forest, mushrooms and a rainbow of jelly fungi—white, orange, red, yellow, purple, indigo—jumped out at me. When I swept aside leaves on the soil surface, a fungal network as bright as neon and regular as a herringbone emerged. I pondered whether fungal networks linked the trees in this tropical forest as they did in my home forests of Douglas fir in British Columbia.

I knew the mycorrhizas in tropical rainforests were mainly the intracellular arbuscular mycorrhizas, not the fleshy extracellular ectomycorrhizas that dominated temperate and boreal forests at higher latitudes. They were generalists, too—a single arbuscular fungal species able to colonize thousands of host plant species, and possibly linking many of the trees and plants together. I also knew there was a multitude of other types of fungi in this forest, from mycorrhizas to saprotrophs, pathogens and endophytes. The prevailing theory of what caused tree patterns and diversity in tropical rainforests was that *pathogenic fungi killed saplings that established near parent trees.* Unlike in the Douglas fir forests, where saplings commonly grew around their parents for protection and shared mutualisms, the saplings in tropical forests often perished from shared pathogens when they grew near parents. This mortality created a patchwork of gaps filled by other tree species.

But the arbuscular mycorrhizal fungi were likely interacting with the pathogenic fungi in complex ways to create the tropical patterns filling my mind. The mycorrhizal fungi could counteract the negative effect of pathogenic fungi and other enemies on tree recruitment, by building up around parent trees and colonizing the emerging saplings. Scientists indeed had found ample evidence that arbuscular mycorrhizal saplings increased in the presence of other arbuscular mycorrhizal trees, and their data suggested that the saplings were facilitated by shared fungal mutualists and potential formation of common mycorrhizal networks with nearby elders. The pathogenic and mycorrhizal fungi appeared to be interacting to help create the patterns and diversity of trees in this tropical rainforest.

On the flight there in a little four-seat Cessna, I'd been amazed to see how intact the forest looked from the air. Achuar territory was a protected Indigenous reserve that had been continuously inhabited by its people for thousands of years. There were no roads, no cities, no clearcuts. I was overwhelmed at seeing the forest in such good condition after coming from British Columbia, where 97 percent of the most productive old-growth forests had been clearcut.

Indigenous-managed territories were some of the most carbon-rich land in the Amazon.[3] But I was aware that portions of the Amazon not protected in Indigenous reserves were imperiled. While 80 percent of the primeval forests—an estimated 390 billion trees, sixteen thousand species—were still here, roads, dams, mines, oil exploration, and cattle ranching were spreading into once-untouched forest. In just the first six months of 2022, more than 3,800 square kilometers (1,500 square miles) of rainforest—three times the size of New York City—had been deforested.[4] Illegal logging was pervasive. Loggers took the valuable hardwoods first, then set fire to the land. The cleared acreage was quickly converted to soybean farms or pasture for the exploding numbers of cattle in the Amazon—more than 80 million head, at last count.

For centuries, the Native peoples of the Amazon countries had been fighting off exploitation of the forests for timber, minerals, fossil fuels, and agriculture. The soils were naturally so highly weathered that removal of the canopy meant instant degradation, and the people depended on intact forests for their livelihoods. The Pachamama Alliance, the sponsor of our trip, had been instrumental in helping to boost the local economy and empower people, especially the women, to more fully participate in the struggle to protect the Amazon. They helped the tribes create alliances to fight global incursions on the rainforest. They provided technical assistance so the local people could launch their own sustainable businesses, such as limited ecotourism—like the lodge where I was staying—and the cultivation of wild vanilla and medicinal plants, creating a cyclical economy.[5]

But it has been the Indigenous people—protecting their traditional knowledge, values, and ways of life—who have used ingenuity, resistance, and guardianship to keep extractive mining, logging, and fossil fuel industries and criminals out of the Amazon rainforest.

The next day, Ramaro took us to one of the most magnificent trees in the Achuar forest: the *samaúma,* or kapok. The crown of the majestic sixty-meter-tall tree emerged from the high canopy like a colossal upside-down umbrella, casting its influence over many hectares of neighbors. The deciduous leaves, capturing the full spectrum of light above the canopy, were photosynthesizing at prodigious rates. The abundance of nooks, grooves, and hollows among the multiple branches hosted a richness of birds, bats, small mammals, and frogs instrumental in dispersing the silky seed, pollen, and nectar of the tree over long distances.

Emergent trees like the kapok, whose crown towers above the canopy, are responsible for about half of the total carbon storage in these forests, and they dominate the energy flow of the ecosystem.[6] These majestic, soaring trees play an outsized role in the functioning and productivity of the forest. Being so much taller than its neighbors means the kapok experiences strong winds. To stabilize itself against lofty air currents, the kapok grows huge, supportive buttresses that can spread up to ten meters from the trunk. The flared roots are mostly above the ground and flow from the trunk like the ruffles on a flamenco skirt. The kapok experiences more drought stress than its neighbors and has a higher risk of cavitation—air bubbles forming in the xylem—in extremely dry periods, which can sometimes kill the tree. The risks of drought stress and cavitation were increasing with climate change, posing a threat to these specialized emergent trees.[7] Just like the big trees in British Columbia. Back home at our project sites, we had begun thinning out the smaller trees beneath—once a function of Indigenous burning—in an attempt to defend these old giants from drought stress.

Ramaro climbed five meters to sit on the knee of the tallest buttress and introduced the tree as his grandfather.

"My grandfather is my spiritual leader. I come to him for advice," he said.

He told us that when he'd been invited to study in the United States, requiring him to leave his wife and children for a year, he'd sought guidance from the tree. His grandfather had advised him to follow this path, and the knowledge Ramaro gained during his year away had eventually helped him to protect the rainforest where he lived.

Broken pieces of pottery were scattered about, and I learned that Ramaro's ancestors were buried in urns among the roots of the grandfather tree. The tree was his conduit to the spirit world of his ancestors. After speaking to us, he sat silently for a long time. When he climbed down from the buttress, he showed us how to lean into the roots and listen. When I tapped on one of the buttresses, the sound echoed through the forest. Ramaro said that the Achuar people and

Ramaro of the Achuar people and Lynne Twist of the Pachamama Alliance at the roots of a kapok, Ramaro's grandfather tree, in the Amazon rainforest in Ecuador, 2023. The Achuar people understand that *arútam*, the spirit of the forest, resides in shaman trees, the giant kapoks, and in waterfalls. Ramaro's ancestors are buried in urns among the roots of the grandfather tree. The nooks, grooves, and hollows in the massive tree are home to an abundance of birds, bats, frogs, and small mammals who help disperse the tree's seeds, pollen, and nectar over long distances.

other Amazonian tribes would use the buttresses of the great trees as drums to communicate with one another. If the roots were anything like those of the Douglas firs to the north, they would spread twice as far as the tree was tall, channeling water and energy back and forth over vast areas. I imagined the water rushing under the bark, surging up the xylem as the stomata exhaled vapor into clouds. The water inched upward with capillary action, and this incredible hydraulic conductivity enabled the trees' great rates of photosynthesis and carbon sequestration.

One hundred meters farther into the forest, Ramaro introduced us to a tree almost as tall as the grandfather and with a burl looking like a child waiting to be born. I could easily imagine that the roots of these two grandparents were linked together.

"We are guided by the spirits in the trees, and with them, we dream of what the future will bring," he said.

We learned that the forest guardian spirit called *arútam* often inhabits the sacred kapok tree. *Arútam* may also appear as a jaguar, a volcano, an anaconda, a waterfall, a bolt of lightning. This supernatural force, no matter what form it takes, is the essence of the forest itself.

As I gazed up at the towering grandmother kapok, I realized I was among kindred spirits. Here in the Amazon, spending time with a people so deeply connected to the natural world, the chasm between Western science and other ways of knowing had never seemed so stark. I wondered what my critics would make of a culture so attuned to the forest that they recognized the trees as kin.

ON MY LAST DAY in the Amazon, I walked back to the great grandmother tree and climbed onto one of her curving buttresses. I leaned back against the tree trunk and let the chattering, trilling, buzzing symphony of the jungle wash over me. I watched as a procession of leafcutter ants carefully skirted around me, descending the trunk carrying their precious cargo of leaf fragments. These tiny ants are essential members of the forest community. By "pruning" leaves in plants and trees, the ants help stimulate new growth and create gaps in the canopy, allowing precious sunlight to penetrate. Every species of plant and animal in the forest community has a role to play, each making its own unique contribution to the perpetuation of the ecosystem. Even, and perhaps especially, people. Humans have always been stewards of the forest, although not always good stewards. Now we need to be protectors, too.

And me, what was my role? I had come to the Amazon feeling sapped in spirit and hoping to find clarity about my next steps in life. I closed my eyes and remembered what Ramaro had said about being guided by the spirits in the trees. Perhaps the tree spirits in this towering matriarch would show me what to do next. As I often reminded my students, there is always more in a forest than what you can see with your own eyes. Perhaps I would have a vision of *arútam*. Or perhaps I would just have a pleasant morning alone in the jungle.

As I sat there with my eyes closed, my body cradled by the sturdy roots of the grandmother tree, I began to imagine that I was a tree, too, at home in the forest of British Columbia. I could almost smell the sharp, earthy fragrance of Douglas fir as I reached out with leaves unfurling to greet the sun. Over time, I grew branches and twigs and more leaves, filling every space to capture more light, building roots that anchored me to the soil. Year after year, I packed on wood in rings, my sapwood expanding with the area of my leaves so I could transmit water from my roots, through my stem, to my prodigious crown of leaves.

Now I tower sixty meters above the ground, weighing hundreds of tons, my wood strong enough to bear gale-force winds and my bark tough enough to douse a flame or pitch out a swarm of beetles. I am a mother tree, the hub of the forest, connected through roots and mycorrhizas to all around me, a key member of a community, countless species working together to make a forest.

As I grow older, I expand my bole more slowly and my crown becomes tattered by the wind. I pulse greater amounts of energy into my social network, feeding the soil, building carbon stocks for future generations, nourishing the communities of saplings growing in the gaps around me. In time, I am spent. I lose my needles and my ability to photosynthesize. As my last needles drop, I become infested with new life, fungi and bugs and bacteria all busily decomposing my great mass. Woodpeckers carve holes in my shedding body, making nests like condos in a high-rise apartment, laying their eggs and raising their young. I have become a snag that is host to a city of heterotrophs, making homes, rotting my bark, and returning me to Mother Earth. A bear makes a den in my hollowing shell, and in spring her cubs emerge from beneath my roots.

Finally, on the heels of a late-spring storm, I fall over. I am now a log, what foresters call coarse woody debris. As my wood softens and decays, a teeming food web of organisms moves in, carrying with them nitrogen and phosphorus, enriching me. Over the decades, I slowly seep into the soil, mineral grains mixing with my organic fibers. A seed falls and germinates on my ancient carcass and I become the seedbed for a new seedling, and the great cycle starts all over again.

I opened my eyes. I recalled the day forty years ago when a grizzly had treed me and Jean in the forest. Perched high in a Douglas fir, clinging to the trunk while I waited for the bear to leave, I had felt part of the tree, melded with its bark. I remembered Titania, the enormous cedar elder at Fairy Creek, shielding me and Nava as we huddled in the protection of her roots. I remembered the comfort I'd gotten from the maple tree at home when I'd been so sick during chemotherapy. The old maple had sheltered me, absorbed me into her heartwood, given me the strength I needed to persevere. *The trees were in my blood. They were my kin.*

I climbed down off the buttress of the kapok tree and headed back to the lodge. It was time to go home.

FLYING FROM ACHUAR TERRITORY to Puyo the next morning, I was glued to the window, enraptured by the river of forest flowing beneath me. On the road winding through the Andes back to Quito, I saw family farms squeezed by commercial plantations of eucalyptus and radiata pine, people's livelihoods encumbered by hasty economic development. I reflected on what the Mother Tree Project had revealed over the past decade. Perhaps my findings could be useful in shining light into the darkness. Perhaps my own experiences with motherhood and loss could give me insight on how to be helpful.

As my flight from Quito approached Vancouver, the forest emerged from the clouds. Air travel and Google Earth had made it impossible to ignore the immensity of loss. I had almost forgotten what a healthy landscape looked like. Visiting the expansive Amazon rainforests in Ecuador had reminded me that the forests of Canada, too, were once vast and healthy.

As the wheels of the aircraft hit the tarmac, I felt decades of loss weighing on my shoulders—not just for the people I'd lost, but for the immense woodlands that once covered the planet. The relentless decline in our forests, caused by burning fossil fuels and logging, was affecting every aspect of our life-support systems: climate regulation, fire regimes, insect and pathogen regimes, carbon storage, water cycles, permafrost, and biodiversity, not to mention our spirits and our cultures. We were in crisis and we needed to be working together to solve these existential problems.

I gathered my bags and stood in line to disembark with the other passengers. People were kind to one another, helping the elderly and the young, all of us weary with jet lag and fatigue. I smiled at the humanity in this little space. We were self-organizing, just as starlings do, to protect ourselves and the flock.

As I walked through the terminal at Vancouver International Airport, I stopped to look at Bill Reid's *Jade Canoe,* a bronze sculpture symbolizing the artist's Haida heritage and the people's interdependence with the land and the sea.[8] The canoe is filled with thirteen characters drawn from Haida mythology. Raven the trickster, a symbol of the capriciousness of nature, holds the steering oar, plotting the course. Crammed into the small canoe together, the passengers bite and claw, uneasily sharing this precious space yet relying on one another for survival on their journey through life.

During the blockade of Lyell Island in 1985, Reid, though ill with Parkinson's, planted himself in the middle of a logging road where he sat whittling a block of wood. When the loggers arrived and saw the celebrated artist blocking their way, they turned and left.[9]

I couldn't help but think of the critics who had told me to stay in my lane, who said my mother tree research crossed a line. Reid had crossed out of his lane. The land defenders who had put their bodies on the line to protect the trees of Fairy Creek had crossed out of their lane, too. Losing Amanda and Mum had hollowed me out so much, and the criticism of my research had wounded me so deeply that I'd actually considered quitting, walking away from my work. But, like the mother tree I'd become in my vision, I understood that my life, too, was a celebration of creation, collaboration, competition, and evolution. Dying, healing, and recovery are constant in nature, and these teachings were all around. I needed to shift out of my paralysis in loss so I could be adaptive, resilient, and helpful. I needed to stand tall in the wind like the kapok. This meant letting the more destructive things roll away like drops of rain, as Teresa had taught me. We are all born to a life full of twists and turns, and I needed to embrace this with grace and grit. I thought about all I had learned about being in reciprocal relationship with the gifts of creation. About taking responsibility for the health of the living webs and cycles of this Earth, our home. *We are all passengers in the same canoe.* I knew now that I would never give up. Consistent with scientific tradition, we published a rebuttal to the critiques.

CONCLUSION

FINDING OUR WAY FORWARD

I have a recurring daydream.

I fly over the rainforests of the Amazon basin to my home forest of British Columbia and to the tree line near the Arctic Circle. With the baton of a conductor—our ancestral wisdom lifting in symphony—the world's fragmented, degraded, and clearcut forests become whole again. Forests damaged by logging, fires, floods, and beetle infestations are restored. Across the Americas, Europe, Africa, and Asia, the trees stand upright and the people take back their lands. From elders to youth, I see a great resurgence of respect for the natural laws. I see the people once again carrying out their ancestral obligations to care for the gifts of creation, honoring our interdependence. I see the excess greenhouse gases travel back into the leaves, down the phloem, into the roots to fuel the soil. Then, as though a thermostat has been turned, global temperatures slowly come down and balance on Earth is restored. It is a dream of renewal and repair, of collaboration and connection and healing.

Our world's actual reality is quite a contrast to my daydream. Deadly floods, boiling heat, and raging wildfires have plunged the planet into scary times. The legacy of greenhouse gases already emitted by fossil fuel burning, land clearing, and wildfire will last decades, even if we stop burning oil and gas today. The Elephant Hill fire alone released 38 million metric tons of greenhouse gases, and emissions from all wildfires in Canada in the same year nearly equaled its emissions from all other forms of energy use. It is sobering that the forests of Canada have been emitting more carbon dioxide from disturbances—fire, insects, logging—than they have absorbed since 2001. Now this is true for almost all of the major forests of the world. Today, wildfires are a major source of all greenhouse gas emissions from the world's forests.

Canada avoided including wildfire emissions in the Paris Agreement because they are the result of "natural disturbances" that are beyond human control. However, it is abundantly clear that all factors that lead to fire are being influenced by humans. The same is true for all of the natural disturbances, including the massive beetle outbreaks, as well as melting permafrost and declining forests. Failure to account for these emissions isn't serving us well, and this is laid bare in steadily increasing greenhouse gas concentrations. Once the emissions are in the atmosphere, it doesn't matter what the source is—lightning- or human-caused fire, insect outbreaks, clearcutting, salvage logging, or extraction and burning of fossil fuels. All of it is contributing to catastrophic climate change.

In the face of these escalating threats, Jean, Teresa, and I set out to innovate stewardship practices informed by Indigenous wisdom that might help safeguard our forests in an uncertain future. By studying the resilience of trees in a shifting environment, we have learned ways to tend them, so that the trees in turn support a thriving community that includes us, too.

We are facing a dystopian threat, but we are also gifted with the stunning opportunity to create transformational change. No human invention can surpass the natural power of photosynthesis. Nature is generously showing us how it works to help mitigate climate change by sequestering excess carbon dioxide and storing it in the ecosystems, including the soil, trees, and plants. This cycle of healing and recovery is constantly occurring. If we can avoid crossing tipping points, our ecosystems will still be resilient enough to recover and maintain their balance. If we take responsible steps, the solutions for mitigation and recovery are still within our reach.

We have learned many things from the Mother Tree Project, and many more discoveries are still to come, but one chief scientific revelation is that the temperate forests we have studied *store at least half of their carbon belowground,* in the roots and soils and underground food webs beneath our feet. Most of this carbon is in dead trees, plants, and animals that have been recycled into organic matter that is almost indestructible, especially when it becomes chemically bound to clay minerals. Some of it is thousands of years old. In fact, globally there is more carbon stored in these ancient underground pools than there is aboveground in the trees plus the atmosphere combined. This knowledge is key to managing and restoring forests and tackling climate change.

Our data clearly shows that *leaving primary and old-growth forests intact,* allowing the trees to grow to old age, as nature intended, is the best way to mitigate greenhouse gas emissions from land-use change. Releasing this carbon pool and disrupting the forest floor—which together constitute well over half the terrestrial carbon budget—will deliver us quickly to chaos.

Clearcutting, we found, is the *worst thing we can do,* and this was true across the entire climate gradient of the Mother Tree Project. Not only did clearcutting result in an immediate pulse of over half of the forest carbon stocks into the atmosphere or into short-lived wood products, it also caused declines in the diversity of forest-dependent species—shrubs, herbs, lichens, mosses, fungi—and hindered the diversity and abundance of regenerating tree seedlings.

Whole tree harvesting, in which trees are delimbed at the landing and the remnant needles, leaves, and branches often burned, exacerbates these ecological losses and causes more intense degradation. In the time we have to act decisively, these losses are irrecoverable.[1]

Even more worrying, we found that *any* harvesting with heavy machinery—the massive feller bunchers, harvesters, skidders, hoe chuckers, yarders, forwarders, delimbers, dozers, and graders—is having a devastating effect on the forest floor. Carbon pools that take thousands of years to build are immediately reduced by 61 percent with the simple act of logging with heavy machinery. Where this carbon goes, we have not yet traced, but estuaries filled with organic materials and silt in areas where the surrounding forests have been clearcut is a telling sign that some of it, without the anchoring of the old trees, erodes into the streams and is being carried out to the rivers leading to the sea.

Other losses of this soil carbon pool come from the scraping of the forest floor and slash into piles for burning after logging, and still more carbon is emitted from the accelerating decomposition of soil organic matter. Scientists estimate that soils, the biggest carbon sink next to oceans, could be emitting 40 percent more carbon by the end of the century, due to increasing drought. Forest floor losses can be reduced by stopping further logging of primary and old-growth forests. Damage to the forest floor in secondary forests can be minimized by careful, skillful hand-falling, using less destructive machinery, or protecting the forest floor during operations. Demechanizing logging would not only be less damaging to the soil, it would also create more jobs and a higher-value product.

Protecting our remaining wild spaces, including the primary forests, wetlands, and prairies, carefully reforesting cut forests with ecologically suitable mixtures of tree and plant species, and restoring lost or damaged forests will be crucial in restoring the terrestrial carbon stocks scientists estimate have dwindled by up to half worldwide due to practices like clearcutting, mining, and dredging for fossil fuels. We cannot continue taking so much from the land. *We need to take much less and leave much more behind.*

Another significant finding from our experiment is that the mother trees—including the old, dying, or dead trees—are vital to biodiversity and forest health. When left standing, the old trees lead the regeneration of the forest, seed in genes for the next generations, shelter the tender seedlings, and nurture the rivers and the salmon. These ancient trees are irreplaceable. They hold in their genes the legacy of climates past—knowledge we need to deal with today's climate crisis.

The perceptive abilities of older trees were powerfully demonstrated in the spruce forests of northern Italy by my colleague Monica Gagliano and Professor Alessandro Chiolerio. While monitoring the bioelectrical impulses of spruce trees during a solar eclipse, they found that the electrical activity of the trees "became significantly more synchronized around the eclipse."[2] The changes occurred in each tree's electrome, which is the electrical signal that is driven when ions cross plant membranes. "The two older trees in the study had a much more pronounced response to the impending eclipse than the young tree.

This suggests older trees may have developed mechanisms to anticipate and respond to such events, similar to their responses to seasonal changes. The scientists also detected bioelectrical waves traveling between the trees. This suggests older trees may transmit their ecological knowledge to younger trees," supporting the concept that forests are cohesive systems. As with our work, their research underscores the importance of protecting elder trees for enhancing ecosystem resilience and potentially archiving and communicating information.

In essence, our findings suggest that the key to our survival lies in the wisdom that comes with age. In old age, a forest reaches its full potential for storing carbon, nurturing biodiversity, and filtering water. Just as young seedlings thrive in the shelter of the old trees, young people can benefit from the knowledge of elders, and younger Western cultures can learn much from age-old cultures. From the Americas to Africa to Australia, Indigenous societies have long been guardians of the natural world, honoring their responsibilities of care for the land, water, and air. Indigenous communities have accumulated an immense body of knowledge, developed, adapted, and inherited over hundreds of centuries. These knowledge systems hold wisdom and principles for protecting the gifts of creation, helping to restore the degraded lands, and mitigating and adapting to climate change. Based on my learnings from and relationships with my Indigenous and non-Indigenous colleagues and allies, as well as the lifetime I have spent in forests, it is my belief that ancient wisdom and spirit hold essential truths to solving many issues related to global change. The solutions start first with respecting one another and all of our relations.

I am deeply encouraged by the growing Land Back movement of communities reclaiming their stolen territories and revitalizing industrialized forest landscapes, as the Kwiakah, Ma'amtagila, Haida, and many First Nations are doing in western British Columbia. Large expanses of their territories have been stripped of the forests and wild salmon runs by logging companies and commercial fish farms over the past century. Now, the people are starting to return to their land. By practicing regenerative forestry to restore their forests, root gardens, clam beds, and fish weirs, they are restoring abundance for future generations. Increasingly, conservation groups, land trusts, and conservancies are recognizing the value of Indigenous land management and helping to facilitate the return of ancestral lands to their original stewards.[3] Some nations are suing in court, others are beneficiaries of donations from private landowners, and others are raising funds to buy their land back. A land trust in the United States is aiding in the return of 12,000 hectares (30,000 acres) to the Penobscot Nation of Maine. The Trust for Public Land bought the land from a timber investor in 2022, and a year later announced plans to transfer ownership and management of the land to the Penobscot, with no use restrictions, once funds are raised to pay off the loan. This has been followed by a 47,097-acre transfer of ancestral land along the lower Klamath River back to the Yurok Tribe in California—the largest collaborative land-back transfer to an Indigenous tribe in U.S. history.

Though long in coming, provincial governments are acknowledging that land-use decisions ought to be informed by the priorities of the First Nations that know the land best. On April 14, 2024, after decades of difficult negotiations, the government of British Columbia formally ratified the Haida Nation's title to the Haida Gwaii archipelago, recognizing the people's claim to their ancestral land. The Rising Tide land agreement ensures the nation greater sovereignty over its own governmental systems and veto power over all land-use decisions. Many details have yet to be sorted, but the agreement is a resounding victory for Aboriginal title and suggests that more power-sharing agreements between First Nations and provincial governments may be in the works.[4]

A year later, in April 2025, the British Columbia Supreme Court dismissed Teal-Jones's $75 million lawsuit against the Province of British Columbia and the Haida Gwaii Management Council.[5] According to forester Keith Moore, who served as a witness for the HGMC, the victory is an indication that times are changing and that corporations should steer clear of trying to thwart conservation efforts.

The forest defenders' devotion to the primeval trees of Fairy Creek has not been in vain. After a tense election in the fall of 2024, narrowly won by British Columbia's left-leaning New Democratic Party, came the good news that Fairy Creek would finally be protected.[6] In a good-faith agreement signed with the British Columbia Green Party, the government pledged it would "move forward to ensure permanent protection of the Fairy Creek watershed" in partnership with the Pacheedaht and Ditidaht First Nations. While the entities hammer out a plan for long-term stewardship, Fairy Creek remains under the protection of a series of temporary, short-term logging deferrals.

Jean, Teresa, and I continue to dream of an expanded Mother Tree Project. In our vision, a network of collaborators would stretch up and down the Pacific Rim, from Alaska to Patagonia, sharing resources and findings and providing education, tools, and support to empower local communities that want to make the transition to regenerative forest stewardship.

The forests themselves are in a constant state of change. Trees in forest communities send, receive, and archive messages about changing circumstances and stresses in order to survive. While they remain rooted through climatic upheavals, they also migrate—slowly—toward safer areas, disseminating seed, all the while communicating, collaborating, and competing with other species so that they can thrive. Elder trees often take the lead in survival tactics, laying down facilitative and competitive footholds and conveying crucial information. Their seedlings rapidly adapt to changing microclimates, slowly mend the soils, and create habitat for other creatures. Our Western societies can, with a shift in our worldview, begin to apply similar principles to be regenerative, adaptive, and resilient.

Over my six-plus decades living and working in the forests of British Columbia, I have observed time and again the ways in which a forest is a diverse and connected community. Each member in this web of life is interdependent with the other members, and these relationships are vital to the health and structure of the forest and its future.

While forests flourish when their connections, cycles, and systems are tended, I've found the same is true of people. In my own personal journey, both as a scientist and as an advocate for forests, I have learned that working together is what makes us strong. A community may be made up of many different people doing different things, each creating and filling their own niche, but the sense of *community and caring* emerges from all of us working together. We can learn this way of being from the forest, to be part of her regenerative nature. We can find the solutions we crave through our kinship and alliances.

The trees are demonstrating the power of collaboration and resurgence. Go into the *nawalakw,* the supernatural forest, and you, too, will feel the wisdom of the trees. When the forest breathes out, you breathe in. When the forest thrives, you thrive. When the forest lives, you live.

YOU CAN FIND *more details on research methods and results from the Mother Tree Project at mothertreeproject.org.*

ACKNOWLEDGMENTS

My parents, Ellen June Simard (née Gardner) and (Peter) Ernest Charles Simard, were both people of the woods, and they brought my sister Robyn, brother Kelly, and me up with great love and respect for the land and waters of the Monashee Mountains. Our parents were children of horse loggers, farmers, and ranchers who were born, lived, and died in the inland rainforests of British Columbia. It was this vital and spiritual connection to the trees, water, and earth that made us people of the forest. As I wrote this book, my story became deeply enriched by my many mentors, colleagues, and students, Indigenous and non-Indigenous. For this reason, this writing is also a reflection of the many people who have walked with me and generously shared their philosophies, wisdom, and knowledge.

During the period of writing this book, I experienced the deaths of my mother and father and my beloved graduate student Amanda Asay. I would never have finished this book without the help of many compassionate and passionate people, who spent time with me, aided in the research, wrote with me, spoke with me, took me to their territories, invited me into ceremony, and taught me their insights.

I am especially grateful to my agent, Doug Abrams at Idea Architects, for embracing the idea of writing a sequel to *Finding the Mother Tree*. Doug has thrown his love, wisdom, and resources into this work because he believes in its promise for bettering the lives of future generations. I am deeply thankful to my writing coach and developmental editor, Tai Moses, who helped bring song to my prose, flow to the passages, wisdom to the teachings, and waves to the arc of the story. I am grateful to Sarah Rainone, editorial director at Idea Architects, who provided crucial guidance and ensured the train never left the tracks. And to Elizabeth Wachtel at Idea Architects for her well of enthusiasm and incredible skills at creating opportunity. I am deeply appreciative to Katherine Vaz, my writing coach for *Finding the Mother Tree*, for helping develop the book proposal and for teaching me so much about the craft of writing.

I'm grateful to executive editor Emily Cunningham at Knopf Doubleday Publishing Group for making sure the pillars of this book stood tall and were woven tightly together. This took time and patience, for which I am thankful. I send deep gratitude to Tiara Sharma, assistant editor at Knopf, for the many production and editorial matters they attended to. To the entire team at Penguin Random House, in the USA, Canada, and the UK, thank you for your skills, networks, and passion in bringing this book to the world stage. To Kim Ingenito and Eleanor Rummell at the Penguin Random House Speakers Bureau, thank you for creating myriad opportunities for me to speak of my work. I am blessed by the talent of my brother-in-law, Bill Heath, who curated, edited, and provided many of the photos in this book. I invite you to explore Bill's cinematography, which helps illustrate some of the concepts in this book, including the salmon forest, the Mother Tree Project, Spirit Bear (Wáwixáxsi), and Haida Gwaii. I am grateful to artist Anneke Rosch for creating the map showing the places mentioned in the narrative.

I benefited from many teachers as I wrote *When the Forest Breathes*. Chiefs, elders, knowledge keepers, professors, scientists, foresters, artists, writers, filmmakers, students, activists, forest defenders, friends, and family all contributed. I am deeply grateful and honored by your generosity in contributing to this work. If I have forgotten to mention your name below, I apologize, but please know that I am thankful.

To my beautiful best friend, Winnifred Jean Roach (née Mather), with whom I have worked, studied, rejoiced, and cried alongside for most of my life: Jean's love for the forest when we first met in 1982 was so contagious that it forever changed my life. Jean is the research manager of the Mother Tree Project, ensuring every grant is written, paper published, student cared for, and that the fieldwork is completed. To my close colleague Dr. Teresa (Sm'hayetsk) Ryan of the Ts'msyen Nation, who has taught me so generously and eloquently about Indigenous ways of seeing the world. Teresa has not only been my teacher, she has prepared opportunities for me to work beside her in Indigenous communities. To the Mother Tree Project team in the Faculty of Forestry and Environmental Stewardship at the University of British Columbia, including communications director Stephanie Troughton, communications coordinator Marie Le Bihan, and social media coordinator Abigail Millson, thank you for bringing the Mother Tree research to the world, tending to all the details of the program, and supporting the entire team. To Azadeh Ardakani for coming into our group to bring a global perspective on Indigenous movements worldwide and for coordinating the Mitacs research cluster for the Kwiakah Nation. To my colleagues Lori Daniels, Jennifer Grenz, Tara Martin, Teresa Ryan, and many others, for the many conversations on how we, as women scientists working in community, navigate our demanding jobs. To my dean, Dr. Robert Kozak, for his support through the many ups and downs with my allies and critics, and to my department head, Richard Hamelin, for giving me a break with my teaching load and always being quick with a laugh. To Chris Rusnak and colleagues for providing the most astute, bedrock guidance.

To my many Indigenous colleagues and friends, from my spirit, thank you. In particular, Lisa White-Kuuyang, Christian White-Kihlguulans Kihl 'Yahda, Barney Edgars, Lia Rosemary Hart-Skiljaadee Jaada K'yaaltsii, Owen Jones-Hltaawii, Marlene Liddle-Kun Kayangas, Sherri Dick-Sgaanjaad, and Jason Alsop-Gaagwiis of the Haida Nation; Kelly Brown, William Housty, and Ron Martin Jr. of the Heiltsuk Nation; Angela Davidson of the Da'naxda'xw Awaetlala Nation; Irv Speck of the Gwawaenuk tribe; hereditary chief David Mungo Knox, Tom Child-Namsgamk'ala, Kaleb Child, and Coreen Child-Yakawilas of the Kwakiutl Nation; hereditary chief Rande Cook-Makwala, hereditary chief K'odi Nelson, Tsastilqualus Ambers Umbas, and Matthew Ambers of the Ma'amtagila Nation; hereditary chief Ernest Alfred of the 'Namgis, Tlowitsis, and Mamalilikulla Nations; hereditary chief Steven Dick, Mike Dick, and Gavin Woodburn of the Kwiakah Nation; Myrle Ballard of the Lake St. Martin First Nation; Elder Bill Jones of the Pacheedaht Nation; Chief Christine Walkem and Dave Walkem of Nlaka'pamux te Secwépemc Nation; and the late Gerry Morigeau of the Ktunaxa Nation.

To my colleagues who wrote the initial proposal to the National Sciences and Engineering Research Council of Canada so that we could initiate the Mother Tree Project, I thank Deborah Bakker, Dirk Brinkman, Les Lavkulich, Bill Mohn, Greg O'Neill, Brian Pickles, Jason Pither, and Jean and Teresa. Brian was especially crucial in crafting the details of the proposal. To my colleagues whose students have collaborated with the Mother Tree Project in various aspects, including Sally Aitken, Younes Alila, Chelsey Armstrong, Matt Betts, Cole Burton, Nicholas Coops, Lori Daniels, Susan Dudley, Shannon Guichon, Rachel Holt, Tara Martin, Dominik Roeser, and Lizzie Wolkovich.

To the many undergraduate and graduate students, postdocs, and research associates at the University of British Columbia who have inspired and helped to carry out the research mentioned in this book since 2015: Amanda Asay, Jeff Chang, Alexia Constantinou, Camille Defrenne, Kiku Dhanwant, Jessica Fostvedt, Thomson Harris, Oliver Heath, Allen Larocque, Dixi Modi, Natalia Mondi, Ari Murphy-Steed, Gabriel Orrego, Alyssa Robinson, Hannah Sachs, Eva Snyder, Hanno Southam, and Bree Summers. To the Mother Tree crew, you have poured your hearts and souls into the fieldwork, lab work, and data analysis, I can't thank you enough: Gaelin Armstrong, Jacob Beauregard, Julia Burkart, Jeff Chang, Abigail Clancy, Dominique Cook, Kaya Fraser, Josh Green, Rachel Green, Thomson Harris, Lia Hart, Oliver Heath, Magali Holt-Lachance, Liam Jones, Danica Law, Bobby Jo Love, Dan Malvin, Erin Miller, Natalia Mondi, Ari Murphy-Steed, Zoe Neudorf, Alyssa Robinson, Curtis Rock, Hannah Sachs, Nava Sachs, Teah Schacter, Suki Simington, Cougar Smith, Kylee Smith, Eva Snyder, Jasmine Tam, Morgan Tien, Joseph Timmermans, Sophie Vanderbanck, Bailey Williams, and Aidan Zickmantel. To my recent postdocs and research associates Monika Gorzelak, Shannon Guichon, and Allen Larocque, thank you.

To the university research forests, forest companies, community forests, and First Nations who collaborated with us in designing, carrying out, and applying the Mother Tree Project: Brinkman and Associates, British Columbia Ministry of Forests, British Columbia Timber Sales, Canadian Forest Service, Canfor Corporation, Carrier Sekanai Nation, Cheakamus Community Forest, Cook's Ferry Indian Band, Cortes Community Forest Cooperative, Council of Haida Nation, Harrop-Proctor Community Forest, Heiltsuk Nation, Interfor Corporation, Kalesnikoff Lumber Company, Katzie Nation, Klahoose Nation, Ktunaxa Nation, Ma'amtagila Nation, Pacific Regeneration Technologies, Roserim Nurseries, Sepwépemc Nation, Sinixt Nation, University of British Columbia Alex Fraser Research Forest, University of British Columbia Malcolm Knapp Research Forest, and the University of Northern British Columbia John

Prince Research Forest. In particular, I thank the following individuals from these organizations: Ionut Aron, Trevor Ball, Heather Beresford, Kathy Coot, Gerald Cordeiro, Caren Dymond, Audrey Ehman, Stephanie Ewen, Nyana Fiddick, William Foster, Sue Grainger, Dexter Hodder, Tyler Hodgkinson, David Isaac, David Jackson, Paul Lawson, Erik Leslie, Erica Lilles, Mark Lombard, Margo Lore, Dennis MacDonald, Deb MacKillop, Hélène Marcoux, Ed Nedokus, Cheryl Power, Glenda Russo, Timo Schreiber, Robert Seaton, Ian Shakespeare, Philip-Edouard Shay, Christian Shears, Jason Stafford, Kari Stuart-Smith, Kori Vernier, Christine Walkem, Dave Walkem, Michaela Waterhouse, Randy Waterhouse, Zach Wright, and Erica Xie.

To the Awi'nakola Foundation and Tree of Life team, our collaborators in science, art, and Indigenous knowledge systems, including Yakawilas Coreen Child, Rande Cook, Kelly Richardson, Teresa Ryan, Stephanie Smith, Paul Walde, Mark Worthing, director of development and operations Britton Jacob-Schram, and their board of directors, including Marilyn Baptiste, Jaskwaan Bedard, Gwanti'lakw Hunt Cranmer, and Maxine Hayman Matilpi.

To David Perry and Teresa Ryan for coauthoring the rebuttal to my critics, "Opinion: Response to questions about common mycorrhizal networks," thank you. My deep gratitude to reviewer Miranda Hart and an anonymous reviewer for improving the article, to *Frontiers in Forests and Global Change* associate editor Verena Griess for overseeing its publication, and to Monika Gorzelak, Erin Miller, Brian Pickles, Jean Roach, and Steph Troughton for your support and reviews.

None of the work described in this book would have happened without the funding and support of several institutions and foundations. These include the British Columbia Ministry of Forests, Canada Foundation for Innovation (CFI), Daughters for Earth, Donner Canadian Foundation, Forest Carbon Initiative (CFI), Forest Enhancement Society of British Columbia (FESBC), ImpactAssets, Jena and Michael King Foundation, Kalliopeia Foundation, Mathematics of Information Technology and Complex Systems (Mitacs), Natural Sciences and Engineering Research Council of Canada (NSERC), NoVo Foundation, One Earth, and University of British Columbia.

Several individuals read and commented on the manuscript and provided crucial feedback and corrections. Thank you to George Asay, Rande

Cook, Bill Heath, Jody Holmes, Rachel Holt, Lynda Mapes, Jean Roach, Teresa Ryan, Hannah Sachs, Nava Sachs, Robyn Simard, Eva Snyder, Dave Walkem, and Lisa White.

To the many people who simply looked out for me, whether in a personal or professional capacity, I can't thank you enough: Rachel Abrams, Janine Benyus, Anne Haigh, Rachel Holt, Denise McInnes, Trish Miller, Beth Rattner, Anneke Rosch, Lonnie Saliken, Deb Thompson, Lynne and Bill Twist, Marissa Van der Vyver, and Azita and James Walton. A special thank-you to Dr. Marcie Jenner, who has helped and continues to help me stand tall and rooted in the most challenging of circumstances.

To my partner, Mary Elizabeth Austin, who has humbly and steadfastly stuck with me through all the ups and downs. To my immediate family, Robyn Simard and Bill Heath, Tiffany Simard and Ron Myers, Marlene Simard, Matthew and Zoe Simard, Kelly Rose Elizabeth Heath and Paul Andrews, Oliver Raven James Heath and Kaya Fraser, and Don Sachs, Deb Summers, and Bree Summers, thank you. And most importantly, to my daughters Hannah Rebekah Sachs and Nava Sophia Sachs, and their partners Bobby Jo Love and Spencer Pearson-Atkins, for always looking out for me, and for devoting your lives to the good of our forests, our home.

I wrote this book to uphold the forests of the world and the people who love, care for, and live in them. It is safe to say the well-being and peace of the entire global ecosystem is interdependent with the health of our forests, and it is my hope that this book contributes to a revolution of care and compassion for these sacred places. *Vive la forêt.*

LIST OF SPECIES

ANIMALS

badger (*Taxidea taxus*)
black bear (*Ursus americanus*)
brown-headed spider monkey (*Ateles fusciceps*)
capuchin monkey (*Cebus aequatorialis*)
capybara (*Hydrochoerus hydrochaeris*)
Columbian ground squirrels (*Urocitellus columbianus*)
common shrew (*Sorex dispar*)
cougar (*Puma concolor*)
coyote (*Canis latrans*)
Ecuadorian purple tarantula (*Avicularia purpurea*)
ermine (*Mustela erminea*)
fisher (*Pekania pennanti*)
grizzly bear (*Ursus arctos horribilis*)
jaguar (*Panthera onca*)
kokanee salmon (*Oncorhynchus nerka*)
mantled howler monkey (*Alouatta palliata aequatorialis*)
marten (*Martes americana*)
moose (*Alces alces*)
mountain goat (*Oreamnos americanus*)
mule deer (*Odocoileus hemionus*)
northwestern salamander (*Ambystoma gracile*)
Pacific herring (*Clupea pallasii*)
pink dolphin (*Inia geoffrensis*)
red-backed vole (*Clethrionomys gapperi*)
red fox (*Vulpes vulpes*)
red squirrel (*Tamiasciurus hudsonicus*)
Rocky Mountain elk (*Cervus elaphus nelsoni*)
salmon (*Oncorhynchus* spp.)
sea lion (*Zalophus californianus*)
sea otter (*Enhydra lutris*)
Sitka black-tailed deer (*Odocoileus hemionus sitkensis*)
southern mountain caribou (*Rangifer tarandus caribou*)
white-tailed deer (*Odocoileus virginianus*)

246   LIST OF SPECIES

BIRDS

ancient murrelet (*Synthliboramphus antiquus*)
bald eagle (*Haliaeetus leucocephalus*)
black-capped chickadee (*Poecile atricapillus*)
burrowing owl (*Athene cunicularia*)
common raven (*Corvus corax*)
marbled murrelet (*Brachyramphus marmoratus*)
northern goshawk (*Accipiter gentilis*)
pileated woodpecker (*Dryocopus pileatus*)
red-breasted nuthatch (*Sitta canadensis*)
red-breasted sapsucker (*Sphyrapicus ruber*)
sandhill crane (*Antigone canadensis*)
scarlet macaw (*Ara macao*)
spotted sandpiper (*Actitis macularius*)
western screech-owl (*Megascops kennicottii*)
whisky jack (*Perisoreus canadensis*)

HERBS

avalanche lily (*Erythronium grandiflorum*)
bear grass (*Xerophyllum tenax*)
bull thistle (*Cirsium vulgare*)
cacao (*Theobroma cacao*)
Canada goldenrod (*Solidago canadensis*)
canary grass (*Phalaris arundinacea*)
chocolate-petaled riceroot (*Fritillaria camschatcensis*)
cloudberry (*Rubus chamaemorus*)
common dandelion (*Taraxacum officinale*)
common horsetail (*Equisetum arvense*)
couch grass (*Elymus repens*)
English ryegrass (*Lolium perenne*)
fairyslipper (*Calypso bulbosa*)
fireweed (*Chamaenerion angustifolium*)
forget-me-not (*Myosotis sylvatica*)
green wintergreen (*Pyrola chlorantha*)
heart-leaved arnica (*Arnica cordifolia*)
heliconia (*Heliconia* spp.)
Hooker's fairybells (*Prosartes hookeri*)
Indian paintbrush (*Castilleja miniata*)
lady fern (*Athyrium filix-femina*)
largeflower fairybells (*Prosartes smithii*)
large-leaved lupine (*Lupinus polyphyllus*)
licorice fern (*Polypodium glycyrrhiza*)
little prince's pine (*Chimaphila menziesii*)
meadow woollyheads (*Psilocarphus elatior*)
monkey-faced orchid (*Dracula simia*)
northern maidenhair fern (*Adiantum pedatum*)
oak fern (*Gymnocarpium dryopteris*)

orchard grass (*Dactylis glomerata*)
Pacific coast strawberry (*Fragaria chiloensis*)
pearly everlasting (*Anaphalis margaritacea*)
pinegrass (*Calamagrostis rubescens*)
pink wintergreen (*Pyrola asarifolia*)
prickly lettuce (*Lactuca serriola*)
rattlesnake plantain (*Goodyera oblongifolia*)
rosy pussytoes (*Antennaria rosea*)
scarlet gilia (*Ipomopsis aggregata*)
single delight (*Moneses uniflora*)
skunk cabbage (*Lysichiton americanum*)
snowdrop (*Galanthus nivalis*)
spiny wood fern (*Dryopteris expansa*)
springbank clover (*Trifolium wormskioldii*)
threeleaf foamflower (*Tiarella trifoliata*)
twinflower (*Linnaea borealis*)
western rattlesnake root (*Nabalus hastatus*)
white hawkweed (*Hieracium albiflorum*)
wild ginger (*Asarum caudatum*)
wild sarsaparilla (*Aralia nudicaulis*)
wild strawberry (*Fragaria virginiana*)
wild vanilla (*Vanilla planifolia*)
yarrow (*Achillea millefolium*)
yellow-flowered silverweed (*Argentina anserina*)
yellow salsify (*Tragopogon dubius*)

FUNGI

golden chanterelle (*Cantharellus cibarius*)
honey mushrooms (*Armillaria ostoyae*)
king bolete (*Boletus edulis*)
morel (*Morchella esculenta*)
true hedgehog (*Hydnum repandum*)

INSECTS

cockroach (*Megaloblatta longipennis*)
Douglas-fir bark beetle (*Dendroctonus pseudotsugae*)
dragonfly (*Sympetrum* ssp.)
leafcutter ant (*Atta* spp.)
mountain pine beetle (*Dendroctonus ponderosae*)
spruce-bark beetle (*Dendroctonus rufipennis*)
swallowtail butterfly (*Papilionidae* family)
western spruce budworm (*Choristoneura occidentalis*)

LICHENS

coral lichen (*Sphaerophorus globosus*)
dog lichen (*Peltigera aphthosa*)

248 LIST OF SPECIES

fairy puke lichen (*Icmadophila ericetorum*)
horsehair lichen (*Bryoria* spp.)
lung lichen (*Lobaria pulmonaria*)
lungwort (*Pulmonaria* spp.)
old-growth specklebelly lichen (*Pseudocyphellaria rainierensis*)
pelt lichen (*Peltigera* spp.)
reindeer cup lichen (*Cladonia rangiferina*)
witch's hair lichen (*Alectoria sarmentosa*)

LIVERWORT

lime-green liverwort (*Marchantia polymorpha*)

MOSSES

bristly haircap (*Polytrichum piliferum*)
broom forkmoss (*Dicranum scoparium*)
electrified cat's-tail moss (*Rhytidiadelphus triquetrus*)
large leafy moss (*Rhizomnium glabrescens*)
Oregon beaked moss (*Kindbergia oregana*)
pipecleaner mosses (*Rhytidiopsis robusta*)
purple horn-tooth (*Ceratodon purpureus*)
red-mouthed leafy moss (*Mnium spinulosum*)
red-stemmed feathermoss (*Pleurozium schreberi*)
slender beaked moss (*Eurhynchium praelongum*)
sphagnum (*Sphagnum* spp.)
step moss (*Hylocomium splendens*)

SHRUBS

Alaska huckleberry (*Vaccinium akaskense*)
baldhip rose (*Rosa gymnocarpa*)
beaked hazelnut (*Corylus cornuta*)
big sagebrush (*Artemisia tridentata*)
black huckleberry (*Vaccinium membranaceum*)
blue elderberry (*Sambucus cerulea*)
blue huckleberry (*Vaccinium deliciosum*)
bog cranberry (*Vaccinium oxycoccos*)
common juniper (*Juniperus communis*)
crowberry (*Empetrum nigrum*)
devil's club (*Oplopanax horridis*)
false azalea (*Rhododendron menziesii*)
grape (*Vitis vinifera*)
highbush cranberry (*Viburnum edule*)
honeysuckle (*Lonicera utahensis*)
Kinnikinnick (*Arctostaphylos uva-ursi*)
lowbush blueberry (*Vaccinium angustifolium*)
oval-leaved blueberry (*Vaccinium ovalifolium*)
penstemon (*Penstemon gracilis*)
red elderberry (*Sambucus racemosa*)

red raspberry (*Rubus idaeus*)
salal (*Gaultheria shallon*)
salmonberry (*Rubus spectabilis*)
Saskatoon berry (*Amelanchier alnifolia*)
Sitka alder (*Alnus viridis* subsp. *sinuata*)
snowberry (*Symphoricarpos albus*)
soopolallie (*Shepherdia canadensis*)
sticky laurel (*Ceanothus velutinus*)
tall Oregon grape (*Mahonia aquifolium*)
thimbleberry (*Rubus parviflorus*)
woolly Labrador tea (*Rhododendron groenlandicum*)

TREES

balsa (*Ochroma pyramidale*)
bigleaf maple (*Acer macrophyllum*)
black cottonwood (*Populus trichocarpa*)
chestnut (*Castanea sativa*)
choke cherry (*Prunus virginiana*)
coastal Douglas fir (*Pseudotsuga menziesii* var. *menziesii*)
crabapple (*Malus fusca*)
Douglas maple (*Acer glabrum*)
Engelmann spruce (*Picea engelmannii*)
eucalyptus (*Eucalyptus globulus*)
grand fir (*Abies grandis*)
Indian plum (*Oemleria cerasiformis*)
interior Douglas fir (*Pseudotsuga menziesii* var. *glauca*)
interior spruce (*Picea engelmannii* x *glauca*)
kapok (*Ceiba pentandra*)
lodgepole pine (*Pinus contorta* var. *latifolia*)
mahogany (*Swietenia macrophylla*)
mountain hemlock (*Tsuga mertensiana*)
Pacific silver fir (*Abies amabilis*)
Pacific yew (*Taxus brevifolia*)
paper birch (*Betula papyrifera*)
ponderosa pine (*Pinus ponderosa*)
Portugal laurel (*Prunus lusitanica*)
Portuguese oak (*Quercus faginea*)
radiata pine (*Pinus radiata*)
red alder (*Alnus rubra*)
Sitka spruce (*Picea sitchensis*)
strangler fig (*Ficus aurea*)
subalpine fir (*Abies lasiocarpa*)
trembling aspen (*Populus tremuloides*)
western hemlock (*Tsuga heterophylla*)
western larch (*Larix occidentalis*)
western redcedar (*Thuja plicata*)
western white pine (*Pinus monticola*)
yellow cedar (*Callitropsis nootkatensis*)

NOTES

2: KINSHIP

1. One hectare (100 meters × 100 meters) is 2.47 acres, about the size of a regulation baseball or rugby field, or London's Trafalgar Square, or a swimming pool big enough to hold twelve blue whales lined up next to one another.
2. World Wildlife Fund, What is forest degradation and why is it bad for people and wildlife?, worldwildlife.org/stories/what-is-forest-degradation-and-why-is-it-bad-for-people-and-wildlife.
3. A fire is generally considered a megafire when it exceeds 100,000 acres, or 40,500 hectares, and burns with high intensity for a long period of time, causing an unusual amount of destruction.

4: KILLER FORESTS

1. Hot lightning strikes, which channel the electrical current for a longer period of time, are thought to have caused up to 90 percent of wildfires in the United States over the past few decades.
2. Portugal's "killer forest," *Politico*, June 19, 2017, politico.eu/article/portugal-fire-eucalyptus-killer-forest/.
3. From April 1 to October 31, 2023, British Columbia experienced 2,245 wildfires, scorching 2.8 million hectares (6.9 million acres) across the province.
4. In the decades before my daughters were born, less than 20,000 hectares (50,000 acres) burned annually. In 2003, 266,000 hectares went up in smoke.
5. In British Columbia, 1.35 million hectares (3.3 million acres) would burn by the end of 2018, accounting for 60 percent of the fire activity across Canada.

5: GARDENS IN THE FOREST

1. While the Indian Act has been amended many times over the years, the legislation has never been entirely repealed; rcaanc-cirnac.gc.ca/eng/1323350306544/1544711580904.
2. Darcy Matthews, University of Victoria, and Dana Lepofsky, Simon Fraser University.
3. First Nations Development Institute, History of Native Food Systems, firstnations.org/wp-content/uploads/publication-attachments/2%20Fact%20Sheet%20History%20of%20Native%20Food%20Systems%20FNDI.pdf.

4. Allen Larocque, Fish, forests, fungi: Soils in the "salmon forests" of British Columbia, https://dx.doi.org/10.14288/1.0413041.

6: THE MIRACLE OF YEW

1. Amanda K. Asay, Suzanne W. Simard, and Susan A. Dudley, Altering neighborhood relatedness and species composition affects interior Douglas-fir size and morphological traits with context-dependent responses, *Frontiers in Ecology and Evolution* 8 (October 2020), https://doi.org/10.3389/fevo.2020.578524.
2. In her paper, Amanda wrote, "We have demonstrated that the kin response in Douglas fir is influenced by the complexity of the environment in which it grows, and this has significant effects on growth, morphology and mycorrhizal fungal colonization that may affect the success and resiliency of regeneration."
3. Arbuscular mycorrhizal fungal species number in the hundreds worldwide, yet are known to colonize more than two hundred thousand—over 80 percent—of all terrestrial plant species. Because of this low host specificity (each fungal species can colonize numerous different plant species), arbuscular mycorrhizas are thought to link many neighboring plants in the forest in a common mycorrhizal network.
4. *Taxus canadensis, T. baccata,* and *T. wallichiana.*
5. The shade would steepen the photosynthetic carbon concentration gradient between cedar and its neighboring plants, and we thought more carbon would flow from carbon-rich cedar to impoverished yew or maple. Previous experiments, including in my lab, had already demonstrated that shade enhanced carbon flow from sunlit neighbors through mycorrhizal networks.

7: SISTER CEDARS

1. South Moresby: A retrospective, *Forestry Chronicle* 66 (June 1990), pubs.cif-ifc.org/doi/pdf/10.5558/tfc66207-3.
2. Nika Collison Jisgang, ed., *Athlii Gwaii: Upholding Haida Law on Lyell Island* (Locarno Press, 2018).

8: CARBON BOMB

1. C. Tattersall Smith, Russell D. Briggs, Inge Stupak, et al., Effects of whole-tree and stem-only clearcutting on forest floor and soil carbon and nutrients in a balsam fir (*Abies balsamea* (L.) Mill.) and red spruce (*Picea rubens* Sarg.) dominated ecosystem, *Forest Ecology and Management* 519 (September 2022), https://doi.org/10.1016/j.foreco.2022.120325.
2. World scientists' warning of a climate emergency, *BioScience* 70, no. 1 (November 2019).

9: PLACE MATTERS

1. S. W. Simard, W. J. Roach, J. Beauregard, et al., Partial retention of legacy trees protects mycorrhizal inoculum potential, biodiversity, and soil resources while promoting natural regeneration of interior Douglas-fir, *Frontiers in Forests and Global Change* 3 (2021): 620436, doi: 10.3389/ffgc.2020.620436.

10: RECIPROCITY

1. Gabriel Orrego (2018). *Western hemlock regeneration on coarse woody debris is facilitated by linkage into a mycorrhizal network in an old-growth forest.* Master of science thesis, University of British Columbia. https://open.library.ubc.ca/media/stream/pdf/24/1.0365590/4.
2. E. Warner, S. C. Cook-Patton, O. T. Lewis, et al., Young mixed planted forests store more carbon than monocultures—a meta-analysis, *Frontiers in Forests and Global Change* 6 (2023): 1226514. doi: 10.3389/ffgc.2023.1226514.

11: THE FOREST DEFENDERS

1. Indigenous Corporate Training, Inc., Hereditary chiefs vs. elected chiefs: What's the difference?, ictinc.ca/blog/hereditary-chiefs-vs-elected-chiefs, May 17, 2021.
2. The War in the Woods was a months-long series of blockades in 1993 protesting the rampant industrial logging of old-growth rainforests in Clayoquot Sound, British Columbia. The demonstrations by an alliance of First Nations and environmental groups led to the permanent protection of half of Clayoquot's intact old-growth forests and the introduction of stricter logging regulations. Until Fairy Creek, it was the largest act of civil disobedience in Canada's history.
3. C. Morton, R. Trenholm, S. Beukema, et al., *Economic valuation of old growth forests on Vancouver Island: Pilot study; Phase 2—Port Renfrew Pilot,* ESSA Technologies Ltd. for the Ancient Forest Alliance, 2021, ancientforestalliance.org/old-growth-economic-report/.
4. Ditidaht, Huu-ay-aht, and Pacheedaht First Nations, Hišuk ma c̓awak Declaration, huuayaht.org/wp-content/uploads/2021/06/declaration-FINAL-signedpdf.pdf.

12: THE FIRST ECOLOGISTS

1. Kyle A. Artelle, Melanie Zurba, Jonaki Bhattacharyya, et al., Supporting resurgent Indigenous-led governance: A nascent mechanism for just and effective conservation, *Biological Conservation* (December 2019), doi.org/10.1016/j.biocon.2019.108284.
2. B.C. First Nation feeds hungry grizzlies 500 salmon carcasses, CBC News, October 4, 2019.
3. British Columbia Truck Loggers Association: https://www.tla.ca/.
4. L. Mo, C. M. Zohner, P. B. Reich, et al., Integrated global assessment of the natural forest carbon potential, *Nature* 624 (2023): 92–101, https://doi.org/10.1038/s41586-023-06723-z.

14: TREES IN OUR BLOOD

1. Awi'nakola (Tree of Life) Foundation: awinakola.com.
2. Stop the spray: Glyphosate, forest monoculture, and fire, *Watershed Sentinel,* September 27, 2023, watershedsentinel.ca/articles/stop-the-spray/.
3. S. W. Simard, T. (S.) L. Ryan, and D. A. Perry, Opinion: Response to questions about common mycorrhizal networks, *Frontiers in Forests and Global Change* 7 (2025): 1512518, doi: 10.3389/ffgc.2024.1512518.

4. "When, in the early 1960s, I brazenly used such words as 'childhood,' 'adolescence,' 'motivation,' 'excitement' and 'mood' I was much criticized. Even worse was my crime of suggesting that chimpanzees had 'personalities.' I was ascribing human characteristics to nonhuman animals and was thus guilty of that worst of ethological sins—anthropomorphism." Jane Goodall, Chimpanzees: Bridging the Gap, in *The Great Ape Project,* eds. Paola Cavalieri and Peter Singer (St. Martin's, 1993), animal-rights-library.com/texts-m/goodall01.htm.

15: BACK TO THE GARDEN

1. Jet Propulsion Laboratory, Dry and wet seasons in the Amazon basin, misr.jpl.nasa.gov/gallery/dry-and-wet-seasons-amazon-basin/.
2. Amazon Aid, The Amazon and water, amazonaid.org/resources/about-the-amazon/the-amazon-and-water/.
3. World Resources Institute, Indigenous forests are some of the Amazon's last carbon sinks, January 6, 2023, wri.org/insights/amazon-carbon-sink-indigenous-forests.
4. Amazon deforestation hits new record in Brazil, *Washington Post,* July 8, 2022, washingtonpost.com/climate-environment/2022/07/08/amazon-rainforest-deforestation-record-climate/.
5. Pachamama Alliance, Bio-entrepreneurship in the Amazon, news.pachamama.org/bio-entrepreneurship-in-the-amazon.
6. H. Ter Steege, N. C. Pitman, D. Sabatier, et al. (2013). Hyperdominance in the Amazonian tree flora. *Science* 342(6156):1243092.
7. Big trees, like the old-growth forests they inhabit, are declining globally, *Mongabay,* January 26, 2012, news.mongabay.com/2012/01/big-trees-like-the-old-growth-forests-they-inhabit-are-declining-globally/.
8. Sams Original Art, *The Jade Canoe,* samsoriginalart.com/en-us/blogs/art/the-jade-canoe?)srsltid=AfmBOo0Ohne5kshuOv24LxpLGK3_A3DvPFasRRAJcLWX01zgLnbDyo5g.
9. Gerald McMaster, *Iljuwas Bill Reid: Life and Work* (Art Canada Institute, 2020), aci-iac.ca/art-books/iljuwas-bill-reid/significance-and-critical-issues/.

CONCLUSION: FINDING OUR WAY FORWARD

1. Climate scientists say we must achieve net-zero carbon emissions by 2050 to prevent the most catastrophic impacts of the climate crisis.
2. "A living collective": Study shows trees synchronize electrical signals during a solar eclipse, *The Conversation,* theconversation.com/a-living-collective-study-shows-trees-synchronise-electrical-signals-during-a-solar-eclipse-255499.
3. Jim Robbins, How returning lands to native tribes is helping protect nature, *Yale Environment 360,* June 3, 2021, e360.yale.edu/features/how-returning-lands-to-native-tribes-is-helping-protect-nature.
4. It's path-breaking: British Columbia's blueprint for decolonization, *The Guardian,* October 9, 2024, theguardian.com/world/2024/oct/09/british-columbia-blueprint-decolonisation.
5. APTN News, B.C. court rules against logging company in Haida Gwaii dispute, April 3, 2025, aptnnews.ca/national-news/b-c-court-rules-against-logging-company-in-haida-gwaii-dispute/.
6. Shannon Waters, B.C. government aims to permanently protect Fairy Creek, *The Narwhal,* December 13, 2024, thenarwhal.ca/bc-fairy-creek-protection-pledge/.

CRITICAL SOURCES

INTRODUCTION: WHEN THE FOREST BREATHES

Aitken, S. N., Yeaman, S., Holliday, J. A., et al. (2008). Adaptation, migration or extirpation: climate change outcomes for tree populations. *Evolutionary Applications* 1: 95–111.

Allen, C. D., Macalady, A. K., Chenchouni, H., et al. (2010). A global overview of drought and heat-induced tree mortality reveals emerging climate change risks for forests. *Forest Ecology and Management* 259: 660–84.

Arnold, C., Atchison, J., and McKnight, A. (2021). Reciprocal relationships with trees: Rekindling Indigenous wellbeing and identity through the Yuin ontology of oneness. *Australian Geographer* 52(2): 131–47.

Ashton, M. S., and Kelty, M. J. (2018). *The Practice of Silviculture: Applied Forest Ecology.* John Wiley & Sons.

Austin, K. G., Baker, J. S., Sohngen, B. L., et al. (2020). The economic costs of planting, preserving, and managing the world's forests to mitigate climate change. *Nature Communications* 11(1): article 1.

Boonman, C. C. F., Serra-Diaz, J. M., Hoeks, S., et al. (2024). More than 17,000 tree species are at risk from rapid global change. *Nature Communications* 15: 166.

Bradley, C. M., Hanson, C. T., and DellaSala, D. A. (2016). Does increased forest protection correspond to higher fire severity in frequent-fire forests of the western United States? *Ecosphere* 7(10).

Byrne, B., Liu, J., Bowman, K. W., et al. (2024). Carbon emissions from the 2023 Canadian wildfires. *Nature* 633: 835–39.

Charnley, S., Fischer, A. P., and Jones, E. T. (2007). Integrating traditional and local ecological knowledge into forest biodiversity conservation in the Pacific Northwest. *Forest Ecology and Management* 246(1): 14–28.

Clason, A. J., Farnell, I., and Lilles, E. B. (2022). Carbon 5–60 years after fire: Planting trees does not compensate for losses in dead wood stores. *Frontiers in Forests and Global Change* 5: 868024.

Dale, V. H., Joyce, L. A., McNulty, S., et al. (2001). Climate change can affect forests by altering the frequency, intensity, duration, and timing of fire, drought, introduced species, insect and pathogen outbreaks, hurricanes, windstorms, ice storms, or landslides. *BioScience* 51: 723–34.

Drever, C. R., Cook-Patton, S. C., Akhter, F., et al. (2021). Natural climate solutions for Canada. *Science Advances* 7(23): eabd6034.

Eiras-Barca, J., Dominguez, F., Yang, Z., et al. (2020). Changes in South American hydroclimate under projected Amazonian deforestation. *Annals of the New York Academy of Sciences* 1472(1): 104–22.

Ellison, A. M., et al. (2005). Loss of foundation species: Consequences for the structure and dynamics of forested ecosystems. *Frontiers in Ecology and the Environment* 3: 479–86.

Google Earth Pro Version 7.3. (2024). Map of British Columbia. Retrieved October 7, 2024. https://earth.google.com/web/.

Gorley, A., and Merkel, G. (2020). A new future for old forests: A strategic review of how British Columbia manages for old forests within its ancient ecosystems. https://www2.gov.bc.ca/assets/gov/farming-natural-resources-and-industry/forestry/stewardship/old-growth-forests/ strategic-review-20200430.pdf.

Heineman, J. L., Sachs, D. L., Mather, W. J., and Simard, S. W. (2010). Investigating the influence of climatic, site, location, and treatment factors on damage to young lodgepole pine in southern British Columbia. *Canadian Journal of Forest Research* 40: 1109–27.

Hessburg, P. F., Miller, C. L., Parks, S. A., et al. (2019). Climate, environment, and disturbance history govern resilience of western North American forests. *Frontiers in Ecology and Evolution* 7: 239.

Hogan, J. A., Domke, G. M., Zhu, K., and Lichstein, J. W. (2024). Climate change determines the sign of productivity trends in US forests. *PNAS* 121(4): e2311132121.

Hogg, E. H., Brandt, J. P., and Krochtubajda, R. (2002). Growth and dieback of aspen forests in northwestern Alberta, Canada, in relation to climate and insects. *Canadian Journal of Forest Research* 32: 823–32.

Holling, C. S. (1973). Resilience and stability of ecological systems. *Annual Review of Ecology, Evolution, and Systematics* 4: 1–23.

International Panel on Climate Change. (2023). *Climate Change 2023: Synthesis Report. Contribution of Working Groups I, II and III to the Sixth Assessment Report of the Intergovernmental Panel on Climate Change.* Core writing team, H. Lee and J. Romero. IPCC. https://doi.org/10.59327/IPCC/AR6-9789291691647.

Ke, P., Ciais, P., Sitch, S., et al. (2024). Low latency carbon budget analysis reveals a large decline of the land carbon sink in 2023. arXiv: 2407.12447.

Levin, S. (2005). Self-organization and the emergence of complexity in ecological systems. *BioScience* 55: 1075–79.

Lindenmayer, D. B., Bowd, E. J., Taylor, C., and Likens, G. E. (2022). The interactions among fire, logging, and climate change have sprung a landscape trap in Victoria's montane ash forests. *Plant Ecology.*

Lindenmayer, D. B., and Laurance, W. F. (2017). The ecology, distribution, conservation and management of large old trees. *Biological Reviews* 92: 1434–58.

Martin, K., and J. M. Eadie. (1999). Nest Webs: A community wide approach to the management and conservation of cavity nesting birds. *Forest Ecology and Management* 115: 243–57.

Mather, W. J., Simard, S. W., Sachs, D. L., and Heineman, J. L. (2010). Decline of planted lodgepole pine in the southern interior of British Columbia. *Forestry Chronicle* 86: 484–97.

Moomaw, W., Yamba, F., Kamimoto, M., et al. (2011). Introduction. In *IPCC Special Report on Renewable Energy Sources and Climate Change Mitigation,* edited by O. Edenhofer, R. Pichs-Madruga, Y. Sokona, et al. Cambridge University Press.

Mueller, R. C., Scudder, C. M., Porter, M. E., et al. (2005). Differential tree mortality in response to severe drought: Evidence for long-term vegetation shifts. *Journal of Ecology* 93: 1085–93.

Newton, A. C. (2021). *Ecosystem Collapse and Recovery.* Cambridge University Press. https://doi.org/10.1017/9781108561105.

Nyland, R. D., Kenefic, L. S., Bohn, K. K., and Stout, S. L. (2016). *Silviculture: Concepts and Applications,* 3rd ed. Waveland Press, Inc.

Perry, D. A., Oren, R., and Hart, S. C. (2008). *Forest Ecosystems,* 2nd ed. Johns Hopkins University Press.

Price, K., Daust, K., and Holt, R. (2023). Conflicting portrayals of remaining old growth: The British Columbia case. *Canadian Journal of Forest Research* 51: 742–52.

Proulx, Annie. (2016). *Barkskins: A Novel.* Scribner.

Roach, W. J., Simard, S. W., Defrenne, C. E., et al. (2021). Tree diversity, site index, and carbon storage decrease with aridity in Douglas-fir forests in western Canada. *Frontiers in Forests Global Change* 4: 682076.

Robinson, J. M., Gellie, N., MacCarthy, D., et al. (2021). Traditional ecological knowledge in restoration ecology: A call to listen deeply, to engage with, and respect Indigenous voices. *Restoration Ecology* 29: e13381.

Simard, S. W. (2018). Mycorrhizal networks facilitate tree communication, learning and memory. In *Memory and Learning in Plants,* edited by F. Baluska, M. Gagliano, and G. Witzany. Springer, 191–213.

Simard, S. W. (2021). *Finding the Mother Tree: Discovering the Wisdom of the Forest.* Knopf.

Simard, S. W., Martin, K., Vyse, A., and Larson, B. (2013). Meta-networks of fungi, fauna and flora as agents of complex adaptive systems. In *Managing World Forests as Complex Adaptive Systems: Building Resilience to the Challenge of Global Change,* edited by K. Puettmann, C. Messier, and K. D. Coates. Routledge, 133–64.

Suding, K. N., Gross, K. L., and Houseman, G. R. (2004). Alternative states and positive feedbacks in restoration ecology. *Trends in Ecology & Evolution* 19: 46–53.

Turner, N. J., Ignace, M. B., and Ignace, R. (2000). Traditional ecological knowledge and wisdom of Aboriginal peoples in British Columbia. *Ecological Applications* 10: 1275–87.

Wang, T., Hamann, A., Spittlehouse, D., and Carroll, C. (2016). Locally downscaled and spatially customizable climate data for historical and future periods for North America. *PLOS One* 11: e0156720.

Woods, A., Coates, K. D., and Hamman, A. (2005). Is an unprecedented Dothistroma needle blight epidemic related to climate change? *BioScience* 55: 761–69.

Zald, H. S., and Dunn, C. J. (2018). Severe fire weather and intensive forest management increase fire severity in a multi-ownership landscape. *Ecological Applications* 28(4): 1068–80.

1: CAN'T GET AWAY FROM YOUR ROOTS

Anderegg, W. R. L., Chegwidden, O. S., Badgley, G., et al. (2022). Future climate risks from stress, insects and fire across US forests. *Ecology Letters* 25(6): 1510–20.

Bärlocher, F., and Rennenberg, H. (2014). Food chains and nutrient cycles. In *Ecological Biochemistry,* edited by G.-J. Krauss and D. H. Nies. Wiley Online Library. https://doi.org/10.1002/9783527686063.ch6.

Betts, M. G., Yang, Z., Hadley, A. S., et al. (2022). Forest degradation drives widespread avian habitat and population declines. *Nature Ecology & Evolution* 6: 709–19.

British Columbia Ministry of Forests and Range and British Columbia Ministry of Environment. (2010). *Field Manual for Describing Terrestrial Ecosystems,* 2nd ed. Forest Science Program, Victoria, B.C. Land 12 Manage. Handb. No. 25. www.for.gov.bc.ca/hfd/pubs/Docs/Lmh/LMH25-2.htm.

Canadian Forest Inventory Committee and Canadian Forest Service. (2008). *Canada's National Forest Inventory: Ground Sampling Guidelines.* Version 5.0. http://www.worldcat.org/title/canadas-national-forest-inventory-ground-sampling-guidelines/oclc/60658012.

Charlie, L. A., and Turner, N. J. (2021). *Luschiim's Plants: Traditional Indigenous Foods, Materials, and Medicines.* Harbour Publishing.

Hagmann, R. K., Hessburg, P. F., Prichard, S. J., et al. (2021). Evidence for widespread changes in the structure, composition, and fire regimes of western North American forests. *Ecological Applications* 31(8): e02431.

Halofsky, J. E., Peterson, D. L., and Harvey, B. J. (2020). Changing wildfire, changing forests: The effects of climate change on fire regimes and vegetation in the Pacific Northwest, USA. *Fire Ecology* 16(1): 4.

Harmon, M. E., Franklin, J. F., Swanson, F. J., et al. (2004). Ecology of coarse woody debris in temperate ecosystems. *Advances in Ecological Research* 34: 59–234.

Hohmann-Marriott, M. F., and Blankenship, R. E. (2011). Evolution of photosynthesis. *Annual Review of Plant Biology* 62: 515–48.

Ingham, R. E., Trofymow, J. A., Ingham, E. R., and Coleman, D. C. (1985). Interactions of bacteria, fungi, and their nematode grazers: Effects on nutrient cycling and plant growth. *Ecological Monographs* 55: 119–40.

Kimmerer, R. W., and Lake, F. K. (2001). The role of indigenous burning in land management. *Journal of Forestry* 99(11): 36.

McDowell, N. G., Beerling, D. J., Breshears, D. D., et al. (2011). The interdependence of mechanisms underlying climate-driven vegetation mortality. *Trends in Ecology and Evolution* 26(10): 523–32.

Mother Tree Project. https://mothertreeproject.org/.

Parish, R., Coupe, R., and Lloyd, D. (1999). *Plants of Southern Interior British Columbia,* 2nd ed. Lone Pine Publishing.

Parisien, M.-A., Barber, Q. E., Bourbonnais, M. L., et al. (2023). Abrupt, climate-induced increase in wildfires in British Columbia since the mid-2000s. *Communications Earth & Environment* 4: 1–11.

Parks, S. A., and Abatzoglou, J. T. (2020). Warmer and Drier Fire Seasons Contribute to Increases in Area Burned at High Severity in Western US Forests from 1985 to 2017. *Geophysical Research Letters* 47(22): e2020GL089858.

Roach, W. J., Simard, S. W., Defrenne, C. E., et al. (2021). Tree diversity, site index, and carbon storage decrease with aridity in Douglas-fir forests in western Canada. *Frontiers in Forests Global Change* 4: 682076.

Roach, W. J., Simard, S. W., and Snyder, E. N. (2024). Downed woody debris varies with climate and harvesting treatment in Douglas-fir forests of British Columbia, Canada. *Frontiers in Forests Global Change* 7: 1397142.

Sayedi, S. S., Abbott, B. W., Vannière, B., et al. (2024). Assessing changes in global fire regimes. *Fire Ecology* 20(1): 18.

Seidl, R., Thom, D., Kautz, M., et al. (2017). Forest disturbances under climate change. *Nature Climate Change* 7: 395–402.

Turner, N. J. (1988). "The Importance of a Rose": Evaluating the Cultural Signifi-

cance of Plants in Thompson and Lillooet Interior Salish. *American Anthropologist* 90(2): 272.
Turner, N. J. (2008). *The Earth's Blanket: Traditional Teachings for Sustainable Living.* University of Washington Press.
Turner, N. J., and Cocksedge, W. (2001). Aboriginal use of non-timber forest products in northwestern North America. *Journal of Sustainable Forestry* 13: 31–58.
Turner, N. J., Łuczaj, Ł. J., Migliorini, P., et al. (2011). Edible and Tended Wild Plants, Traditional Ecological Knowledge and Agroecology. *Critical Reviews in Plant Sciences* 30(1–2): 198–225.
van Nes, E. H., Scheer, M., Brovkin, V., et al. (2015). Causal feedbacks in climate change. *Nature Climate Change* 5: 445–48.
Veraverbeke, S., Verstraeten, W. W., Lhermitte, S., et al. (2012). Assessment of post-fire changes in land surface temperature and surface albedo, and their relation with fire—Burn severity using multitemporal MODIS imagery. *International Journal of Wildland Fire* 21(3): 243.
Wikipedia. St. Joseph's Mission (Williams Lake). https://en.wikipedia.org/wiki/Saint _Joseph%27s_Mission_(Williams_Lake).
Williams Lake First Nation. St. Joseph's Mission Investigation. https://www.wlfn.ca /about-wlfn/sjm-investigation/sjm-investigation-releases/.

2: KINSHIP

Anderegg, W. R., Schwalm, C., Biondi, F., et al. (2015). Pervasive drought legacies in forest ecosystems and their implications for carbon cycle models. *Science* 349(6247): 528–32.
Asay, A. K. (2013). *Mycorrhizal facilitation of kin recognition in interior Douglas-fir (*Pseudotsuga menziesii *var.* glauca*).* Master of science thesis, University of British Columbia.
Asay, A. K. (2019). *Influence of kin, density, soil inoculum potential and interspecific competition on interior Douglas-fir (*Pseudotsuga menziesii *var.* glauca*) performance and adaptive traits.* PhD dissertation, University of British Columbia.
Beiler, K. J., Durall, D. M., Simard, S. W., et al. (2010). Mapping the wood-wide web: Mycorrhizal networks link multiple Douglas-fir cohorts. *New Phytologist* 185: 543–53.
Beiler, K. J., Simard, S. W., and Durall, D. M. (2015). Topology of *Rhizopogon* spp. mycorrhizal meta-networks in xeric and mesic old-growth interior Douglas-fir forests. *Journal of Ecology* 103: 616–28.
Beiler, K. J., Simard, S. W., Lemay, V., and Durall, D. M. (2012). Vertical partitioning between sister species of *Rhizopogon* fungi on mesic and xeric sites in an interior Douglas-fir forest. *Molecular Ecology* 21: 6163–74.
British Columbia Forest Practices Board. (2023). *Forest and fire management in BC: Toward landscape resilience.* SR 61. https://www.bcfpb.ca/wp-content/uploads/2023 /06/SR61-Landscape-Fire-Management.pdf.
Buotte, P. C., Law, B. E, Ripple, W. J., and Berner, L. T. (2019). Carbon sequestration and biodiversity co-benefits of preserving forests in the western United States. *Ecological Applications* 30(2).
Buotte, P. C., Levis, S., Law, B. E., et al. (2019). Near-future forest vulnerability to drought and fire varies across the western United States. *Global Change Biology* 25: 290–303.

Caverley, N. (2008). Understanding the human dimensions of the mountain pine beetle infestation: Lessons learned from the First Nations Mountain Pine Beetle Initiative. In *Mountain Pine Beetle: From Lessons Learned to Community-based Solutions.* Conference proceedings, June 10–11, 2008. *BC Journal of Ecosystems and Management* 9(3): 150–53.

Chen, J., Chen, W., Liu, J., et al. (2000). Annual carbon balance of Canada's forests during 1895–1996. *Global Biogeochemical Cycles* 14(3): 839–49.

Christianson, A. C., Sutherland, C. R., Moola, F., et al. (2022). Centering Indigenous voices: The role of fire in the boreal forest of North America. *Current Forestry Reports* 8(3): 257–76.

Cook's Ferry Indian Band. https://cooksferry.ca/.

DellaSala, D. A., Baker, B. C., Hanson, C. T., et al. (2022). Have western USA fire suppression and megafire active management approaches become a contemporary Sisyphus? *Biological Conservation* 268: 109499.

Dellert, L. H. (1998). Sustained Yield: Why Has It Failed to Achieve Sustainability? in *The Wealth of Forests: Markets, Regulation, and Sustainable Forestry,* edited by C. Tollefson. University of British Columbia Press, 255–77. https://doi.org/10.59962/9780774852425-013.

Dudley, S. A., and File, A. L. (2007). Kin recognition in an annual plant. *Biology Letters* 3: 435–38.

Early, S. K. (2024). Deadwood: People, place, and neoliberal forest policy in British Columbia, Canada. *Nature and Space* 7(1): 288–310.

Ellison, D., Morris, C. E., Locatelli, B., et al. (2017). Trees, forests and water: Cool insights for a hot world. *Global Environmental Change* 43: 51–61.

File, A. L., Klironomos, J., Maherali, H., and Dudley, S. A. (2012). Plant kin recognition enhances abundance of symbiotic microbial partner. *PLOS One* 7: e45648.

Franklin, J. F., Cromack, K., Jr., Denison, W., et al. (1981). *Ecological Characteristics of Old-Growth Douglas-Fir Forests.* Gen. Tech. Rep. PNW-GTR-118. U.S. Department of Agriculture, Forest Service, Pacific Northwest Forest and Range Experiment Station. https://doi.org/10.2737/PNW-GTR-118.

Gorzelak, M. A. (2017). *Kin-selected signal transfer through mycorrhizal networks in Douglas-fir.* PhD dissertation, University of British Columbia. https://doi.org/10.14288/1.0355225.

Gorzelak, M., Pickles, B. J., Asay, A. K., and Simard, S. W. (2015). Inter-plant communication through mycorrhizal networks mediates complex adaptive behaviour in plant communities. *Annals of Botany Plants* 7: plv050.

Government of Canada (2004). A Guide to Understanding the Canadian Environmental Protection Act, 1999. Canada.ca. https://www.canada.ca/en/environment-climate-change/services/canadian-environmental-protection-act-registry/publications/guide-to-understanding.html.

Hagerman, S. M., Dowlatabadi, H., and Satterfield, T. (2010). Observations on drivers and dynamics of environmental policy change: Insights from 150 years of forest management in British Columbia. *Ecology and Society* 15(1).

Hogan, J. A., Domke, G. M., Zhu, K., et al. (2024). Climate change determines the sign of productivity trends in US forests. *Proceedings of the National Academy of Sciences* 121(4): p.e2311132121.

Ignace, M., and Ignace, R. E. (2017). *Secwépemc People, Land, and Laws: Yerí7 re stsq'ey's-kucw,* vol. 90. McGill-Queen's University Press.

Isabel, N., Halliday, J., and Aitken, S. N. (2019). Forest genomics: Advancing climate adaptation, forest health, productivity, and conservation. *Evolutionary Applications* 13: 3–10.

Johnstone, J. F., Allen, C. D., Franklin, J. F., et al. (2016). Changing disturbance regimes, ecological memory, and forest resilience. *Frontiers in Ecology and the Environment* 14: 369–78.

Korkiakoski, M., Tuovinen, J. P., Penttilä, T., et al. (2019). Greenhouse gas and energy fluxes in a boreal peatland forest after clear-cutting. *Biogeosciences* 16(19): 3703–23.

Kuraś, P. K., Alila, Y., and Weiler, M. (2012). Forest harvesting effects on the magnitude and frequency of peak flows can increase with return period. *Water Resources Research* 48(1).

Kurz, W. A., Dymond, C. C., Stinson, G., et al. (2008). Mountain pine beetle and forest carbon feedback to climate change. *Nature* 452(7190): 987–90.

Kurz, W. A., Shaw, C. H., Boisvenue, C., et al. (2013). Carbon in Canada's boreal forest—A synthesis. *Environmental Reviews* 21(4): 260–92.

Law, B. E., and Harmon, M. E. (2011). Forest sector carbon management, measurement and verification, and discussion of policy related to climate change. *Carbon Management* 2(1): 73–84.

Lesmeister, D. B., Davis, R. J., Singleton, P. H., and Wiens, J. D. (2018). Northern spotted owl habitat and populations: Status and threats. In *Synthesis of Science to Inform Land Management Within the Northwest Forest Plan Area*, edited by T. A. Spies, P. A. Stine, R. Gravenmier, et al. Gen. Tech. Rep. PNW-GTR-966. U.S. Department of Agriculture, Forest Service, Pacific Northwest Research Station, 245–99.

Leverkus, A. B., Buma, B., Wagenbrenner, J., et al. (2021). Tamm review: Does salvage logging mitigate subsequent forest disturbances? *Forest Ecology and Management* 481: 118721.

Levine, J. I., Collins, B. M., Steel, Z. L., et al. (2022). Higher incidence of high-severity fire in and near industrially managed forests. *Frontiers in Ecology and the Environment* 20(7): 397–404.

Lindenmayer, D. B., Burton, P. J., and Franklin, J. F. (2012). *Salvage Logging and Its Ecological Consequences.* Island Press.

Liu, Q., Peng, C., Schneider, R., et al. (2023). Drought-induced increase in tree mortality and corresponding decrease in the carbon sink capacity of Canada's boreal forests from 1970 to 2020. *Global Change Biology* 29(8): 2274–85.

Lloyd, D., Angove, K., Hope, G., and Thompson, C. (1990). *A Guide to Site Identification and Interpretation for the Kamloops Forest Region.* Land Management Handbook 23. British Columbia Ministry of Forests.

Lochhead, K. D., Kleynhans, E. J., and Muhly, T. B. (2022). Linking woodland caribou abundance to forestry disturbance in southern British Columbia, Canada. *The Journal of Wildlife Management* 86(1): p.e22149.

Luyssaert, S., Schulze, E.-D., Börner, A., et al. (2008). Old-growth forests as global carbon sinks. *Nature* 455 (7211): 213–15.

MacKillop, D., Ehman, A., Iverson, K., and MacKenzie, E. (2018). *A Field Guide to Ecosystem Classification and Identification for Southeast British Columbia.* Land Management Handbook 71. B.C. Ministry of Forests, Lands, Natural Resource Operations and Rural Development.

Marchak, M. P., Aycock, S. L., and Herbert, D. M. (1999). *Falldown: Forest Policy in British Columbia.* David Suzuki Foundation.

Meidinger, D. V., and Pojar, J. (compilers and editors). (1991). Ecosystems of British

Columbia. B.C. Ministry of Forests, Research Branch, Victoria, B.C. Special Report Series 6.

Metsaranta, J. M., Shaw, C. H., Kurz, W. A., et al. (2017). Uncertainty of inventory-based estimates of the carbon dynamics of Canada's managed forest (1990–2014). *Canadian Journal of Forest Research* 47(8): 1082–94.

Moomaw, W. R., and Law, B. E. (2023). A call to reduce the carbon costs of forest harvest. *Nature* 620: 44–45.

Nlaka'pamux Nation Tribal Council. https://nntc.ca/.

Pan, Y., Birdsey, R. A., Phillips, O. L., et al. (2024). The enduring world forest carbon sink. *Nature* 631(8021): 563–69.

Peng, L., Searchinger, T. D., Zionts, J., and Waite, R. (2023). The carbon costs of global wood harvests. *Nature* 620: 110–15.

Perry, D. A., Oren, R., and Hart, S. C. (2008). *Forest Ecosystems,* 2nd ed. Johns Hopkins University Press.

Pham, H. C., and Alila, Y. (2023). Science of forests and floods: The quantum leap forward needed, literally and metaphorically. *Science of the Total Environment* 912: 169646.

Pickles, B. J., Wilhelm, R., Asay, A. K., et al. (2017). Transfer of 13C between paired Douglas-fir seedlings reveals plant kinship effects and uptake of exudates by ectomycorrhizas. *New Phytologist* 214: 400–11.

Porter, B., Gregovich, D. P., Crupi, A. P., et al. (2021). Black bears select large woody structures for dens in southeast Alaska. *The Journal of Wildlife Management* 85(7): 1450–61.

Pousette, J., and Hawkins, C. (2006). An assessment of critical assumptions supporting the timber supply modelling for mountain-pine-beetle-induced allowable annual cut uplift in the Prince George Timber Supply Area. *BC Journal of Ecosystems and Management* 7(2): 93–104.

Simard, S. W. (2009). The foundational role of mycorrhizal networks in self-organization of interior Douglas-fir forests. *Forest Ecology and Management* 258: S95—S107.

Simard, S. W. (2017). The mother tree. In *The Word for World Is Still Forest,* edited by A.-S. Springer and E. Turpin. K. Verlag & the Haus der Kulturen der Welt.

Simard, S. W., Jones, M. D., Durall, D. M., and Sorensen, N. S. (1998). Regeneration of a dry, grassy site in the Interior Douglas-fir zone. In *Managing the Dry Douglas-Fir Forests of the Southern Interior: Workshop Proceedings,* edited by A. Vyse, C. Hollstedt, and D. Huggard. BC Working Paper Vol. 34. British Columbia Ministry of Forests.

Simard, S. W., Jones, M. D., Durall, D. M., et al. (2003). Chemical and mechanical site preparation: Effects on Pinus contorta growth, physiology, and microsite quality on grassy, steep forest sites in British Columbia. *Canadian Journal of Forest Research* 33(8): 1495–515.

Steen, O., and Coupe, R. A. (1997). *A Field Guide to Forest Site Identification and Interpretation for the Cariboo Forest Region.* Land Management Handbook 39. British Columbia Ministry of Forests.

Stuwix Resources Joint Venture. Welcome to Stuwix. https://www.stuwix.com; Thomas, G. (2008). "Beetle Beat": A First Nations' perspective on how the First Nations people have been affected by the mountain pine beetle. In *Mountain Pine Beetle: From Lessons Learned to Community-Based Solutions.* Conference proceedings, June 10–11, 2008. *BC Journal of Ecosystems and Management* 9(3): 60–64.

Thonfeld, F., Gessner, U., Holzwarth, S., et al. (2022). A first assessment of canopy cover loss in Germany's forests after the 2018–2020 drought years. *Remote Sensing* 14: 562.

Van Dorp, C., Simard, S. W., and Durall, D. M. (2020). Resilience of Rhizopogon-Douglas-fir mycorrhizal networks 25 years after selective logging. *Mycorrhiza* 30: 467–74.
Waring, R. H., and Running, S. W. (2007). *Forest Ecosystems,* 3rd ed. Academic Press.

3: PLANTING SEASON

Aitken, S. N., Yeaman, S., Holliday, J. A., et al. (2008). Adaptation, migration or extirpation: Climate change outcomes for tree populations. *Evolutionary Applications* 1: 95–111.
Almendras, K., García, J., Carú, M., and Orlando, J. (2018). Nitrogen-fixing bacteria associated with *Peltigera* cyanolichens and *Cladonia* chlorolichens. *Molecules* 23: 3077.
Ashton, M. S., and Kelty, M. J. (2018). *The Practice of Silviculture: Applied Forest Ecology.* John Wiley & Sons.
Bardgett, R. D., and Van Der Putten, W. H. (2014). Belowground biodiversity and ecosystem functioning. *Nature* 515(7528): 505–11.
British Columbia Ministry of Forests. (2023). *2020 Major Timber Processing Facilities in British Columbia.* Government of British Columbia. https://www2.gov.bc.ca/gov/content/industry/forestry/competitive-forest-industry/forest-industry-economics/fibremill-information/major-timber-processing-facilities-survey.
Bysouth, D., Boan, J. J., Malcolm, J. R., and Taylor, A. R. (2024). High emissions or carbon neutral? Inclusion of "anthropogenic" forest sinks leads to underreporting of forestry emissions. *Frontiers in Forests Global Change* 6: 1297301.
Crowther, T. W., Van den Hoogen, J., Wan, J., et al. (2019). The global soil community and its influence on biogeochemistry. *Science* 365(6455): p.eaav0550.
Eberhard, S., Finazzi, G., and Wollman, F. A. (2008). The dynamics of photosynthesis. *Annual Review of Genetics* 42(1): 463–515.
Environment and Climate Change Canada (2024). *National Inventory Report 1990–2022: Greenhouse Gas Sources and Sinks in Canada. Canada's Submission to the United Nations Framework Convention on Climate Change,* part 1. Government of Canada.
Gies, E. (2023). *Water Always Wins: Thriving in an Age of Drought and Deluge.* University of Chicago Press.
Grenz, J. (2024). *Medicine Wheel for the Planet: A Journey Toward Personal and Ecological Healing.* University of Minnesota Press.
Harris, T. C., Roach, W. J., Miller, E. M., and Simard, S. W. (2025). The interactive role of climatic transfer distance and overstory retention on douglas fir seedling survival and height growth in interior British Columbia. *Global Change Biology* 31(1): e70027.
Ingham, R. E., Trofymow, J. A., Ingham, E. R., and Coleman, D. C. (1985). Interactions of bacteria, fungi, and their nematode grazers: Effects on nutrient cycling and plant growth. *Ecological Monographs* 55: 119–40.
Isaac-Renton, M. G., Roberts, D. R., Hamann, A., and Spiecker, H. (2014). Douglas-fir plantations in Europe: A retrospective test of assisted migration to address climate change. *Global Change Biology* 20: 2607–17.
Lehmann, J., and Kleber, M. (2015). The contentious nature of soil organic matter. *Nature* 528: 60–68.
Mann, M. E. (2007). Climate over the past two millennia. *Annual Review of Earth and Planetary Sciences* 35: 111–36.

Marshall, D. J. (2024). Principles of experimental design for ecology and evolution. *Ecology Letters* 27(4): p.e14400.

Mclellan, B. (2023). *Grizzly Bear Science and the Art of a Wilderness Life: Forty Years of Research in the Flathead Valley.* Rocky Mountain Books.

Nyland, R. D. (2016). *Silviculture Concepts and Applications,* 3rd ed. Waveland Press, Inc.

O'Neill, G. A., Stoehr, M., and Jaquish, B. (2014). Quantifying safe seed transfer distance and impacts of tree breeding on adaptation. *Forest Ecology and Management* 328: 122–30.

Parish, R., Coupe, R., Lloyd, D. (1999). *Plants of Southern Interior British Columbia,* 2nd ed. Lone Pine Publishing.

Polanyi, M., Skene, J., and Simard, A. (2024). *Reported Greenhouse Gas (GHG) Emissions from Logging in Canada Double After Revision to Government Data.* Nature Canada, NRDC, Nature Quebec. https://naturecanada.ca/wp-content/uploads/2024/09/2024-Logging-Emissions-Update-Report.pdf.

Prescott, C. E., and Grayston, S. J. (2023). TAMM review: Continuous root forestry—Living roots sustain the belowground ecosystem and soil carbon in managed forests. *Forest Ecology and Management* 532: 120848.

Puettmann, K. J., Coates, K. D., and Messier, C. C. (2012). *A Critique of Silviculture: Managing for Complexity.* Island Press.

Rehfeldt, G. E., Jaquish, B. C., López-Upton, J., et al. (2014). Comparative genetic responses to climate for the varieties of *Pinus ponderosa* and *Pseudotsuga menziesii*: Realized climate niches. *Forest Ecology and Management* 324: 126–37.

Roach, W. J. (2019). Designing successful forest renewal practices for our changing climate: The Mother Tree Project—Project update. doc_project-summary_09-2019_1.pdf.

Sáenz-Romero, C., O'Neill, G., Aitken, S. N., and Lindig-Cisneros, R. (2020). Assisted migration field tests in Canada and Mexico: Lessons, limitations, and challenges. *Forests* 12: 9.

Selosse, M. A., Bocayuva, M. F., Kasuya, M. C. M., and Courty, P. E. (2016). Mixotrophy in mycorrhizal plants: Extracting carbon from mycorrhizal networks. *Molecular Mycorrhizal Symbiosis*: 451–71.

Simard, S. W., Roach, W. J., Beauregard, J., et al. (2021). Partial retention of legacy trees protects mycorrhizal inoculum potential, biodiversity and soil resources while promoting natural regeneration of interior Douglas-fir. *Frontiers in Forests and Global Change* 3: 620436.

Simard, S. W., Roach, W. J., Defrenne, C. E., et al. (2020). Harvest intensity effects on carbon stocks and biodiversity are dependent on regional climate in Douglas-fir forests of British Columbia. *Frontiers in Forests and Global Change* 3: 88.

Snyder, E. N., Guichon, S., and Simard, S. W. (2025). We care about yew: Marker gene sequencing of arbuscular mycorrhizal fungal communities associated with western yew, Douglas maple and western redcedar. *Nature Scientific Reports,* submitted.

Terborgh, J. (2020). At 50, Janzen–Connell has come of age. *BioScience* 70(12): 1082–92.

Trainor, K. (2020). Yellow Glacier Lily, *Erythronium grandiflorum*. *Anthropocenes–Human, Inhuman, Posthuman* 1: 16.

Varhola, A., Coops, N. C., Weiler, M., and Moore, R. D. (2010). Forest canopy effects on snow accumulation and ablation: An integrative review of empirical results. *Journal of Hydrology* 392: 219–33.

Varner, C. (2022). *Invasive Flora of the West Coast: British Columbia and the Pacific Northwest.* Heritage House Publishing Co.

Wang, H., Bi, X., and Clift, R. (2021). Utilization of forestry waste materials in British Columbia: Options and strategies. *Renewable and Sustainable Energy Reviews* 150: 111480.
Zahra, S., Novotny, V., and Fayle, T. M. (2021). Do reverse Jansen-Connell effects reduce diversity? *Trends in Ecology and Evolution* 36: 387–90.

4: KILLER FORESTS

Agee, J. K. (1993). Fire ecology of Pacific Northwest forests. Island Press.
Ames, P. (2017). Portugal's "killer forest." *Politico.* https://www.politico.eu/article/portugal-fire-eucalyptus-killer-forest/.
Baron, J. N., Gergel, S. E., Hessburg, P. F., and Daniels, L. D. (2022). A century of transformation: Fire regime transitions from 1919 to 2019 in southeastern British Columbia, Canada. *Landscape Ecology* 37(10): 2707–27.
Bowd, E. J., Banks, S. C., Strong, C. L., and Lindenmayer, D. B. (2019). Long-term impacts of wildfire and logging on forest soils. *Nature Geoscience* 12(2): 113–18.
Bowman, D. M., Balch, J., Artaxo, P., et al. (2011). The human dimension of fire regimes on Earth. *Journal of Biogeography* 38(12): 2223–36.
British Columbia Forest Practices Board. (2023). *Forest and Fire Management in BC: Toward Landscape Resilience.* Special Report 61. https://www.bcfpb.ca/wp-content/uploads/2023/06/SR61-Landscape-Fire-Management.pdf.
British Columbia Ministry of Environment, Lands and Parks. (1996). Floodplain mapping: Bonaparte River at Cache Creek. https://www.env.gov.bc.ca/wsd/data_searches/fpm/reports/bc-floodplain-design-briefs/bonaparte_river_cache_creek.pdf.
British Columbia Ministry of Forests. (2004). *2003 Firestorm Provincial Review. Forest Protection Program.* British Columbia Ministry of Forests. https://www2.gov.bc.ca/assets/gov/public-safety-and-emergency-services/wildfire-status/governance/bcws_protection_submission_to_filmon.pdf.
Brookes, W., Daniels, L. D., Copes-Gerbitz, K., et al. (2021). A disrupted historical fire regime in central British Columbia. *Frontiers in Ecology and Evolution* 9: 676961.
Byrne, B., Liu, J., Bowman, K. W., et al. (2024). Carbon emissions from the 2023 Canadian wildfires. *Nature* 633: 835–39.
Canadian Broadcasting Corporation. (2015). B.C. wildfire in 2003 remembered by woman who lost her home. CBC News. https://www.cbc.ca/news/canada/british-columbia/b-c-wildfire-in-2003-remembered-by-woman-who-lost-her-home-1.3142465.
Canadian Broadcasting Corporation. (2015). Province-wide wildfires blanket B.C. communities in smoke. CBC News. https://www.cbc.ca/news/canada/british-columbia/province-wide-wildfires-blanket-b-c-communities-in-smoke-1.3138999.
Canadian Broadcasting Corporation. (2020). One of B.C.'s most destructive wildfires was most likely caused by a smoker, investigations find. CBC News. https://www.cbc.ca/news/canada/british-columbia/elephant-hill-wildfire-cause-investigation-smoking-1.5554355.
Canadian Broadcasting Corporation. (2023). Kelowna residents relive stories of Okanagan Mountain Park fire on its 20th anniversary. CBC News. https://www.cbc.ca/news/canada/british-columbia/okanagan-mountaino-park-fire-20-anniversary-1.6937590.
Charlbois, B. (2023). Ongoing costs from Elephant Hill wildfire pegged at $1B a year: Report. Globalnews.ca.
Clason, A. J., Farnell, I., and Lilles, E. B. (2022). Carbon 5–60 years after fire: Planting

trees does not compensate for losses in dead wood stores. *Frontiers in Forests Global Change* 5: 868024. doi.org/10.3389/ffgc.2022.868024.

Colleta, A., Mooney, C., Dennis, B., et al. (2022). Canada's Elephant Hill fire burned for 3 months. No country claims its greenhouse gas emissions. *Washington Post.*

Davis, K. T., Dobrowski, S. Z., Higuera, P. E., et al. (2019). Wildfires and climate change push low-elevation forests across a critical climate threshold for tree regeneration. *Proceedings of the National Academy of Sciences* 116(13): 6193–98.

Davis, K. T., Dobrowski, S. Z., Holden, Z. A., et al. (2019). Microclimatic buffering in forests of the future: The role of local water balance. *Ecography* 42(1): article 1. https://doi.org/10.1111/ecog.03836.

DeBano, L. F. (2000). The role of fire and soil heating on water repellency in wildland environments: A review. *Journal of Hydrology* 231: 195–206.

Dickson-Hoyle, S., and Char, J. (2021). *Elephant Hill: Secwépemc leadership and lessons learned from the collective story of wildfire recovery.* Secwepemcúl'ecw Restoration and Stewardship Society.

Ellison, D., Morris, C. E., Locatelli, B., et al. (2017). Trees, forests and water: Cool insights for a hot world. *Global Environmental Change* 43: 51–61.

Frayer, L. (2017). Reeling from its deadliest forest fire, Portugal finds a villain: Eucalyptus trees. *Los Angeles Times.* www.latimes.com/world/europe/la-fg-portugal-eucalyptus-fire-20170620-story.html.

Gies, E. (2024). As logging intensifies forest fires, Wet'suwet'en fight to protect old growth. *Mongabay.* https://news.mongabay.com/2024/09/as-logging-intensifies-forest-fires-wetsuweten-fight-to-protect-old-growth/.

Government of British Columbia. (2023). Wildfire Averages. June 15, 2023. https://www2.gov.bc.ca/gov/content/safety/wildfire-status/about-bcws/wildfire-statistics/wildfire-averages.

Hagerman, S. M., Dowlatabadi, H., and Satterfield, T. (2010). Observations on drivers and dynamics of environmental policy change: Insights from 150 years of forest management in British Columbia. *Ecology and Society* 15(1).

Hall, E. (2010). *Maintaining Fire in British Columbia's Ecosystems: An Ecological Perspective.* https://sernbc.ca/uploads/library/additional_related/fire_fuel_management/Maintaining_Fire_in_BC_Ecosystems.pdf.

Hanes, C. C., Wang, X., Jain, P., et al. (2019). Fire-regime changes in Canada over the last half century. *Canadian Journal of Forest Research* 49: 256–69.

Hessburg, P. F., Agee, J. K., and Franklin, J. F. (2005). Dry forests and wildland fires of the inland Northwest USA: Contrasting the landscape ecology of the pre-settlement and modern eras. *Forest Ecology and Management* 211: 117–39.

Hessburg, P. F., Miller, C. L., Parks, S. A., et al. (2019). Climate, environment, and disturbance history govern resilience of western North American Forests. *Frontiers in Ecology and Evolution* 7: 239.

Hessburg, P. F., Spies, T. A., Perry, D. A., et al. (2016). Tamm review: Management of mixed-severity fire regime forests in Oregon, Washington, and Northern California. *Forest Ecology and Management* 366: 221–50.

Hood, S., Abrahamson, I., and Cansler, C. A. (2018). Fire resistance and regeneration characteristics of Northern Rockies tree species. In *Fire Effects Information System.* U.S. Department of Agriculture, Forest Service, Rocky Mountain Research Station, Missoula Fire Sciences Laboratory. https://www.fs.fed.us/database/feis/pdfs/other/FireResistRegen.html.

Howe, M., Peng, L., and Carroll, A. (2022). Landscape predictions of western balsam bark beetle activity implicate warm temperatures, a longer growing season, and drought in widespread irruptions across British Columbia. *Forest Ecology and Management* 508: 120047.

Jain, P., Barber, Q. E., Taylor, S. W., et al. (2024). Drivers and impacts of the record-breaking 2023 wildfire season in Canada. *Nature Communications* 15(1): 6764.

Johnson, R. S., and Alila, Y. (2023). Nonstationary stochastic paired watershed approach: Investigating forest harvesting effects on floods in two large, nested, and snow-dominated watersheds in British Columbia, Canada. *Journal of Hydrology* 625: 129970.

Jorgenson, M. (2024). Wild smoke: Managing forest pollution in northern British Columbia since 1950. *Environment and History* 30: 267–90.

Kimmerer, R. W., and Lake, F. K. (2001). The role of Indigenous burning in land management. *Journal of Forestry* 99: 36–41.

Kline, J. R., Reed, K. L., Waring, R. H., and Stewart, M. L. (1976). Field measurement of transpiration in Douglas-fir. *Journal of Applied Ecology* 13: 273–83.

Kurz, W. A., Dymond, C. C., Stinson, G., et al. (2008). Mountain pine beetle and forest carbon: Feedback to climate change. *Nature* 452: 987–90.

Liang, L., Liang, S., and Zeng, Z. (2024). Extreme climate sparks record boreal wildfires and carbon surge in 2023. *The Innovation* 5(4): 100631.

Lindenmayer, D. B., Bowd, E. J., Taylor, C., and Likens, G. E. (2022). The interactions among fire, logging, and climate change have sprung a landscape trap in Victoria's montane ash forests. *Plant Ecology* 223: 733–49.

Lindenmayer, D. B., Yebra, M., and Cary, G. J. (2023). Perspectives: Better managing fire in flammable tree plantations. *Forest Ecology and Management* 528: 120641.

Liu, Z., Ballantyne, A. P., and Cooper, L. A. (2019). Biophysical feedback of global forest fires on surface temperature. *Nature Communications* 10(1): 214.

Morman, K. E., Buckley, H. L., Higgins, C. M., et al. (2024). Simulated fire and plant-soil feedback effects on mycorrhizal fungi and invasive plants, *iScience* 27(11): 111193.

Parisien, M. A., Barber, Q. E., Flannigan, M. D., and Jain, P. (2023). Broadleaf tree phenology and springtime wildfire occurrence in boreal Canada. *Global Change Biology* 29(21): 6106–19.

Pearce, F. (2017). Rivers in the Sky: How Deforestation Is Affecting Global Water Cycles. *Yale Environment 360*.

Pérez-Invernón, F. J., Gordillo-Vázquez, F. J., Huntrieser, H., and Jöckel, P. (2023). Variation of lightning-ignited wildfire patterns under climate change. *Nature Communications* 14(1): 739.

Pham, H. C., and Alila, Y. (2023). Science of forests and floods: The quantum leap forward needed, literally and metaphorically. *Science of the Total Environment* 912: 169646.

Pilz, D., McLain, R., Alexander, S., et al. (2007). Ecology and management of morels harvested from the forests of western North America. Gen. Tech. Rep. PNW-GTR-710. U.S. Department of Agriculture, Forest Service, Pacific Northwest Research Station. https://research.fs.usda.gov/treesearch/26906.

Popović, Z., Bojović, S., Marković, M., and Cerdà, A. (2021). Tree species flammability based on plant traits: A synthesis. *Science of the Total Environment* 800: 149625.

Prescott, C. E., Blevins, L. L., and Staley, C. L. (2000). Effects of clear-cutting on decomposition rates of litter and forest floor in forests of British Columbia. *Canadian Journal of Forest Research* 30: 1751–57.

Roden, B. (2015). May flood in Cache Creek similar to 1979, especially in Valleyview. *The Ashcroft-Cache Creek Journal.* https://www.ashcroftcachecreekjournal.com/news/may-flood-in-cache-creek-similar-to-1979-especially-in-valleyview-5826254.

Shrestha, S., Williams, C. A., Rogers, B. M., et al. (2022). Wildfire controls on land surface properties in mixed conifer and ponderosa pine forests of Sierra Nevada and Klamath mountains, Western US. *Agricultural and Forest Meteorology* 320: 108939.

Simard, S. W., Hagerman, S. M., Sachs, D. L., et al. (2005). Conifer growth, *Armillaria ostoyae* root disease and plant diversity responses to broadleaf competition reduction in temperate mixed forests of southern interior British Columbia. *Canadian Journal of Forest Research* 35: 843–59.

Simard, S. W., and Vyse, A. (2006). Trade-offs between competition and facilitation: A case study of vegetation management in the interior cedar-hemlock forests of southern British Columbia. *Canadian Journal of Forest Research* 36: 2486–96.

Smith, W. J., Bekessy, S. A., Ward, M., and Wintle, B. A. (2024). Dealing with the risk of fire in carbon sequestration strategies: Diverse forests or plantation monocultures? *Conservation Science and Practice:* e13201.

Veraverbeke, S., Rogers, B. M., Goulden, M. L., et al. (2017). Lightning as a major driver of recent large fire years in North American boreal forests. *Nature Climate Change* 7(7): 529–34.

Vose, J. M., Peterson, D. L., Domke, G. M., et al. (2018): Forests. In *Impacts, Risks, and Adaptation in the United States: Fourth National Climate Assessment,* vol. 2, edited by D. R. Reidmiller, C. W. Avery, D. R. Easterling, et al. U.S. Global Change Research Program, 232–67. https://doi.org/10.7930/NCA4.2018.CH6.

Winkler, R. D., Spittlehouse, D. L., and Golding, D. L. (2005). Measured differences in snow accumulation and melt among clearcut, juvenile, and mature forests in southern British Columbia. *Hydrological Processes: An International Journal* 19: 51–62.

Woo, H., Bone, C., Nadeem, K., and Taylor, S. W. (2024). Influence of mountain pine beetle outbreaks on large fires in British Columbia. *Ecosphere* 15: e4722.

Woods, A. J., Heppner, D., Kope, H. H., et al. (2010). Forest health and climate change: A British Columbia perspective. *The Forestry Chronicle* 86: 412–22.

Zald, H. S., and Dunn, C. J. (2018). Severe fire weather and intensive forest management increase fire severity in a multi-ownership landscape. *Ecological Applications* 28(4): 1068–80.

5: GARDENS IN THE FOREST

Armstrong, C. G., Grenz, J., Zyp-Loring, J., et al. (2024). Ethnoecological perspectives on environmental stewardship: Tenets and basis of reciprocity in Gitxsan and nłeʔkepmx (Nlaka'pamux) Territories. *People and Nature* 7: 934–46.

Armstrong, C. G., Lyons, N., McAlvay, A. C., et al. (2023). Historical ecology of forest garden management in Laxyuubm Ts'msyen and beyond. *Ecosystems and People* 19(1): 2160823.

Armstrong, C. G., Miller, J., McAlvay, A. C., et al. (2021). Historical indigenous land-use explains plant functional trait diversity. *Ecology and Society* 26(2): 6.

Atlas, W. I., Ban, N. C., Moore, J. W., et al. (2021). Indigenous systems of management for culturally and ecologically resilient Pacific salmon (*Oncorhynchus* spp.) fisheries. *BioScience* 71(2): 186–204.

Coastal First Nations Great Bear Initiative. (2000). Land-based Healing at Kunsoot.

Coastal First Nations. https://coastalfirstnations.ca/resources/land-based-healing-at-kunsoot/.

Connaughton, S. P., Hill, G., Morin, J., et al. (2022). Tidal belongings: First nations—Driven archaeology to preserve a large wooden fish trap panel recovered from the Comox Harbour Intertidal Fish Trap Complex in British Columbia. *Canadian Journal of Archaeology* 46(1): 16–51.

Deur, D., Dick, A., Recalma-Clutesi, K., and Turner, N. J. (2015). Kwakwaka'wakw "clam gardens" motive and agency in traditional northwest coast mariculture. *Human Ecology* 43: 201–12.

Groesbeck, A. S., Rowell, K., Lepofsky, D., and Salomon, A. K. (2014). Ancient clam gardens increased shellfish production: Adaptive strategies from the past can inform food security today. *PLOS One* 9(3): p.e91235.

Harding, J. N., and Reynolds, J. D. (2014). Opposing forces: Evaluating multiple ecological roles of Pacific salmon in coastal stream ecosystems. *Ecosphere* 5: art157.

Heiltsuk Nation. (2012–2017). Húy̓at. https://www.hauyat.ca/home.html.

Helfield, J. M., and Naiman, R. J. (2006). Keystone interactions: Salmon and bear in riparian forests of Alaska. *Ecosystems* 9: 167–80.

Hocking, M. D., and Reynolds, J. D. (2011). Impacts of salmon on riparian plant diversity. *Science* 331 (6024): 1609–12.

Hopetown Community Development Society. (2024). Guardians. https://hcds.ca/index.php/guardians/.

Jones, J. T. (1990). *"We looked after all the salmon streams," Traditional Heiltsuk salmon stewardship of salmon and salmon streams: A preliminary assessment.* Master of Arts in the School of Environmental Studies, University of Victoria, Victoria, Canada. https://hdl.handle.net/1828/21335.

Larocque, A. T. (2022). *Fish, forests, fungi: Soils in the "salmon forests" of British Columbia.* Doctoral dissertation, University of British Columbia.

Larocque, A. T., and Simard, S. W. (2023). Legacy of salmon-derived nutrients on riparian soil chemistry and soil fertility on the Central Coast of British Columbia, Canada. *Frontiers in Forests and Global Change* 6: 010294. https://open.library.ubc.ca/soa/cIRcle/collections/ubctheses/24/items/1.0413041.

Lawson, K. D. (2024). *Revitalizing and future proofing Haiłzaqv λiƛábacʼi (clam gardens) through Ǧvi̓ḷás (ancestral laws) du łáxvaí (authority).* Master of Resource Management in the School of Resource and Environmental Management Faculty of Environment, Simon Fraser University, Vancouver, Canada. https://summit.sfu.ca/_flysystem/fedora/2024-09/etd23260.pdf.

Lepofsky, D., Armstrong, C. G., Greening, S., et al. (2017). Historical ecology of cultural keystone places of the Northwest Coast. *American Anthropologist* 119(3): 448–63.

Lepofsky, D., Smith, N. F., Cardinal, N., et al. (2015). Ancient shellfish mariculture on the Northwest Coast of North America. *American Antiquity* 80(2): 236–59.

McCaig, S. (1989). Shuswap River drives. Enderby and District Museum and Archives. https://enderbymuseum.ca/river.php.

McIlwraith, T. (2015). The Splatsin Cooke Creek Culture Camp and the ironies of access to the Shuswap River. *Canadian Journal of Native Studies* 35(2): 153–81.

Museum of Anthropology at the University of British Columbia. (2020). Knowledge Keepers: Cedar Harvest (YouTube). htttps://www.youtube.com/watch?v=A6KS4J8QyNQ.

Quinn, T. P., Carlson, S. M., Gende, S. M., and Rich, H. B., Jr. (2009). Transportation of

Pacific salmon carcasses from streams to riparian forests by bears. *Canadian Journal of Zoology* 87(3): 195–203.

Reimchen, T., and Fox, C. H. (2013). Fine-scale spatiotemporal influences of salmon on growth and nitrogen signatures of Sitka spruce tree rings. *BMC Ecology* 13: 1–13.

Ryan, T. (S.) L. (2013). Summary Comparison of Governance Models That Consider Aboriginal Traditional Knowledge. In *Indigenous Earth: Praxis and Transformations*, edited by E. Simmons. Theytus Books. http://hdl.handle.net/2429/46515.

Ryan, T. (S.) L. (2014). *Territorial jurisdiction: The cultural and economic significance of eulachon* Thaleichthys pacificus *in the north-central coast region of British Columbia.* PhD dissertation, University of British Columbia.

Turner, N. J., Ari, Y., Berkes, F., et al. (2009). Cultural management of living trees: An international perspective. *Journal of Ethnobiology* 29(2): 237–70.

White, E., Moss, M., and Cannon, A. (2011). Heiltsuk stone fish traps on the central coast of British Columbia. In *The Archaeology of North Pacific Fisheries*, edited by Madonna L. Moss and Aubrey Cannon. University of Alaska Press, 75–90.

Williams, M., Stallard, T., and Zalasiewicz, J. (2023). Cosmos. In *Handbook of the Anthropocene: Humans Between Heritage and Future.* Springer International Publishing, 27–31.

6: THE MIRACLE OF YEW

Armstrong, C. G., Clemente-Carvalho, R. B., Turner, N. J., et al. (2024). Genetic differentiation and precolonial Indigenous cultivation of hazelnut (*Corylus cornuta*, Betulaceae) in Western North America. *Proceedings of the National Academy of Sciences* 121(48): e2402304121.

Armstrong, C. G., Dixon, W. C. M., and Turner, N. J. (2018). Management and traditional production of Beaked Hazelnut (k'áp'xw-az', *Corylus cornuta;* Betulaceae) in British Columbia. *Human Ecology* 46: 547–59.

Arsenault, A., and Bradfield, G. E. (1995). Structural compositional variation in three age-classes of temperate rainforests in southern coastal British Columbia. *Canadian Journal of Botany* 73(1): 54–64.

Asay, A. K. (2013). *Mycorrhizal facilitation of kin recognition in interior Douglas-fir (*Pseudotsuga menziesii var. glauca*).* Master of science thesis, University of British Columbia. https://dx.doi.org/10.14288/1.0103374.

Asay, A. K. (2019). *Influence of kin, density, soil inoculum potential and interspecific competition on interior Douglas-fir (*Pseudotsuga menziesii var. glauca*) performance and adaptive traits.* PhD dissertation, University of British Columbia. https://dx.doi.org/10.14288/1.0387438.

Asay, A. K., Simard, S. W., and Dudley, S. A. (2020). Altering neighborhood relatedness and species composition affects interior Douglas-fir size and morphological traits with context-dependent responses. *Frontiers in Ecology and Evolution* 8: 578524.

Bailey, J. D., and Liegel, L. H. (1997). Response of Pacific yew (*Taxus brevifolia*) to partial removal of the overstory. *Western Journal of Applied Forestry* 12(2): 41–43.

Bingham, M. A., and Simard, S. W. (2011). Do mycorrhizal network benefits to survival and growth of interior Douglas-fir seedlings increase with soil moisture stress? *Ecology and Evolution* 3: 306–16.

Bingham, M. A., and Simard, S. W. (2012). Ectomycorrhizal networks of old *Pseudotsuga menziesii* var. *glauca* trees facilitate establishment of conspecific seedlings under drought. *Ecosystems* 15: 188–99.

Busing, R. T., Halpern, C. B., and Spies, T. A. (1995). Ecology of Pacific Yew (*Taxus brevifolia*) in Western Oregon and Washington. *Conservation Biology* 9(5): 1199–1207.

Campbell, E., and Nicholson, A. (1995). *A Summary of Western Yew Biology with Recommendations for Its Management in British Columbia.* Province of British Columbia Ministry of Forests Research Program.

Canadian Broadcasting Corporation. (2021). How some children at the Kamloops residential school died. (YouTube) https://www.youtube.com/watch?v=fP9oHOpo8Mc.

Canadian Broadcasting Corporation. (2022). Kamloops Indian Residential School survivor shares harrowing experience. (YouTube) https://www.youtube.com/watch?v=JA8FIsxnWYI.

Crawford, R. C., and Johnson, F. D. (1985). Pacific yew dominance in tall forests, a classification dilemma. *Canadian Journal of Botany* 63(3): 592–602.

The Fifth Estate. (2022). Kamloops residential school survivors recall students going missing, digging of graves in orchard. (YouTube) https://www.youtube.com /watch?v=m8wXExEHiS8.

Fisher, J. A., Shackelford, N., Hocking, M. D., et al. (2019). Indigenous peoples' habitation history drives present-day forest biodiversity in British Columbia's coastal temperate rainforest. *People and Nature* 1(1): 103–14.

Forrester, D. I., and Bauhus, J. (2016). A review of processes behind diversity—Productivity relationships in forests. *Current Forestry Reports* 2: 45–61.

Gond, S. K., Kharwar, R. N., and White, J. F. (2014). Will fungi be the new source of the blockbuster drug Taxol? *Fungal Biology Reviews* 28(4): 77–84.

Halter, M. R., and Chanway, C. P. (1993). Growth and root morphology of planted and naturally-regenerated Douglas fir and lodgepole pine. *Annales des Sciences Forestières* 50: 71–77.

Heinig, U., Scholz, S., and Jennewein, S. (2013). Getting to the bottom of Taxol biosynthesis by fungi. *Fungal Divers.* 60(1): 161–70.

Hotte, N., Wyatt, S., and Kozak, R. (2019). Influences on trust during collaborative forest governance: A case study from Haida Gwaii. *Canadian Journal of Forest Research* 49(4): 361–74.

Itokawa, H. (2003). *Taxus: The Genus Taxus.* Taylor & Francis.

Kelsey, R. G., and Vance, N. C. (1992). Taxol and cephalomannine concentrations in the foliage and bark of shade-grown and sun-exposed *Taxus brevifolia* trees. *Journal of Natural Products* 55(7): 912–17.

Krawec, P. (2022). *Becoming Kin: An Indigenous Call to Unforgetting the Past and Reimagining Our Future.* Augsburg Fortress Publishers.

LeMay, V., Pommerening, A., and Marshall, P. (2009). Spatio-temporal structure of multi-storied, multi-aged interior Douglas fir (*Pseudotsuga menziesii* var. *glauca*) stands. *Journal of Ecology* 97(5): 1062–74.

MacKillop, D., Ehman, A., Iverson, K., and MacKenzie, E. (2018). *A Field Guide to Ecosystem Classification and Identification for Southeast British Columbia.* Land Management Handbook 71. B.C. Ministry of Forests, Lands, Natural Resource Operations and Rural Development.

Mitchell, A. K. (1998). Acclimation of Pacific yew (*Taxus brevifolia*) foliage to sun and shade. *Tree Physiology* 18(11): 749–57.

Newsome, T. A., Heineman, J. L., Nemec, A. F. L., et al. (2010). Ten-year regeneration responses to varying levels of overstory retention in two productive southern British Columbia ecosystems. *Forest Ecology and Management* 260(1): 132–45.

Nicolaou, K. C., Guy, R. K., and Potier, P. (1996). Taxoids: New weapons against cancer. *Scientific American* 274(6): 94–98.

Pezzuto, J. (1996). Taxol production in plant cell culture comes of age. *Nature* 14: 1083.

Pickles, B. J., Twieg, B. D., O'Neill, G. A., et al. (2015). Local adaptation in migrated interior Douglas-fir seedlings is mediated by ectomycorrhizas and other soil factors. *New Phytologist* 207(3): 858–71.

Pickles, B. J., Wilhelm, R., Asay, A. K., et al. (2017). Transfer of 13C between paired Douglas-fir seedlings reveals plant kinship effects and uptake of exudates by ectomycorrhizas. *New Phytologist* 214: 400–11.

Reynolds, G. E. M. (2022). *Ecological understandings of Indigenous landscape management shape the study of Pacific yew (*Taxus brevifolia*).* Master of arts thesis, University of Victoria.

Simard, S. W. (2009). The foundational role of mycorrhizal networks in self-organization of interior Douglas-fir forests. *Forest Ecology and Management* 258: 95–107.

Simard, S. W. (2012). Mycorrhizal networks and seedling establishment in Douglas-fir forests. In *Biocomplexity of Plant–Fungal Interactions,* edited by Darlene Southworth. Wiley-Blackwell, 85–107.

Simard, S. W., Beiler, K. J., Bingham, M. A., et al. (2012). Mycorrhizal networks: Mechanisms, ecology and modelling. *Fungal Biology Reviews* 26: 39–60.

Simard, S. W., Jones, M. D., Durall, D. M., et al. (1997). Reciprocal transfer of carbon isotopes between ectomycorrhizal *Betula papyrifera* and *Pseudotsuga menziesii*. *New Phytologist* 137: 529–42.

Simard, S. W., Perry, D. A., Jones, M. D., et al. (1997). Net transfer of carbon between tree species with shared ectomycorrhizal fungi. *Nature* 388: 579–82.

Snyder, E. N., Guichon, S., and Simard, S. W. (2025). We care about yew: Marker gene sequencing of arbuscular mycorrhizal fungal communities associated with western yew, Douglas maple and western redcedar. *Nature Scientific Reports,* submitted.

Spies, T. A. (2004). Ecological concepts and diversity of old-growth forests. *Journal of Forestry* 102(3): 14–20.

Stierle, A., Strobel, G., Stierle, D., et al. (1995). The search for a Taxol-producing microorganism among the endophytic fungi of the Pacific yew, *Taxus brevifolia*. *Journal of Natural Products* 58(9): 1315–24.

Stoehr, M., Bird, K., Nigh, G., et al. (2010). Realized genetic gains in coastal Douglas-fir in British Columbia: Implications for growth and yield projections. *Silvae Genetica* 59(1–6): 223–33.

Thomas, P. (2013). The IUCN Red List of Threatened Species. https://www.iucnredlist.org/en.

Turner, N. J. (1998). *Plant Technology of First Peoples in British Columbia.* University of British Columbia Press.

Turner, N. J. (2008). *The Earth's Blanket: Traditional Teachings for Sustainable Living.* D&M Publishers.

Turner, N. J., and Cocksedge, W. (2001). Aboriginal use of non-timber forest products in northwestern North America: Applications and issues. *Journal of Sustainable Forestry* 13: 31–58.

Vyse, A., Ferguson, C., Simard, S. W., et al. (2006). Growth of Douglas-fir, lodgepole pine, and ponderosa pine seedlings underplanted in a partially-cut, dry Douglas-fir stand in south-central British Columbia. *The Forestry Chronicle* 82(5): 723–32.

Vyse, A., Smith, R. A., and Bondar, B. G. (1991). Management of Interior Douglas-fir stands in British Columbia: Past present and future. In *Interior Douglas-fir: The*

Species and Its Management, edited by D. M. Baumgartner. Washington State University, Department of Natural Resource Sciences, 177–85.
Weaver, B. A. (2014). How Taxol/paclitaxel kills cancer cells. *Molecular Biology Cell* 25(18): 2677–81.
Woods, A. J., and Bergerud, W. A. (2008). *Are Free-growing Stands Meeting Timber Productivity Expectations in the Lakes Timber Supply Area?* FREP Report 16. Ministry of Forests and Range, Forest and Range Evaluation Program.
Woods, A., and Coates, K. D. (2013). Are biotic disturbance agents challenging basic tenets of growth and yield and sustainable forest management? *Forestry* 86(5): 543–54.
Wubet, T., Weiß, M., Kottke, I., and Oberwinkler, F. (2003). Morphology and molecular diversity of arbuscular mycorrhizal fungi in wild and cultivated yew (*Taxus baccata*). *Canadian Journal of Botany* 81(3): 255–66.
Zhang, Y., Chen, H. Y., and Reich, P. B. (2012). Forest productivity increases with evenness, species richness and trait variation: A global meta-analysis. *Journal of Ecology* 100: 742–49.

7: SISTER CEDARS

Antos, J. A., Filipescu, C. N., and Negrave, R. W. (2016). Ecology of western redcedar (*Thuja plicata*): Implications for management of a high-value multiple-use resource. *Forest Ecology and Management* 375: 211–22.
Arriagada, C., Manquel, D., Cornejo, P., et al. (2012). Effects of the co-inoculation with saprobe and mycorrhizal fungi on *Vaccinium corymbosum* growth and some soil enzymatic activities. *Journal of Soil Science and Plant Nutrition* 12: 283–94.
Ballantyne Project. (2021). Canadian Residential Schools: The Survivors and Their Descendants. https://www.youtube.com/watch?v=W_sVD6-W1zg.
Banner, A., MacKenzie, W. H., Pojar, J., et al. (2014). *A Field Guide to Ecosystem Classification and Identification for Haida Gwaii.* Land Management Handbook 68. Province of British Columbia.
British Columbia, Environment and Land-use Committee. (1978). South Moresby Island Wilderness Proposal: An Overview. Government of British Columbia. https://www.env.gov.bc.ca/wld/documents/moresby.pdf.
Canadian Broadcasting Corporation. (2023). Stolen totem pole formally welcomed home to Nisga'a territory after nearly a century in Scottish museum. News, September 28, 2023. http://cbc.ca/news/canada/british-columbia/memorial-totem-pole-returned-1.6981891.
Collinson, L. (Chief of Skidegate, 1881–1970). (1966). Address by Chief of Skidegate in March 1966. https://www.ravencallingproductions.ca/chiefskidegatespeech.
Collison Jisgang, N., ed. (2018). *Athlii Gwaii: Upholding Haida Law on Lyell Island.* Locarno Press.
Daghino, S., Martino, E., Voyron, S., and Perotto, S. (2022). Metabarcoding of fungal assemblages in *Vaccinium myrtillus* endosphere suggests colonization of aboveground organs by some ericoid mycorrhizal and DSE fungi. *Scientific Reports* 12(1): 11013.
Editorial Board. (1990). South Moresby: A retrospective. *Forestry Chronicle* 66: 207–8. pubs.cif-ifc.org/doi/pdf/10.5558/tfc66207-3.
Gardner, J. E. (1994). Building alliances to protect South Moresby Island. In *Sea Has Many Voices: Oceans Policy for a Complex World,* edited by C. Lamson. McGill-Queen's University Press, 163–83. https://doi.org/10.1515/9780773564251-012.

Holt, R. F. (2005). *Environmental Conditions Report for the Haida Gwaii/Queen Charlotte Islands Land Use Plan.* Veridian Ecological Consulting. www.bearconservation.org.uk/HaidaGwaiienvcondreport.pdf.

Huntley, M. J., Mathewes, R. W., and Shotyk, W. (2013). High-resolution palynology, climate change and human impact on a late Holocene peat bog on Haida Gwaii, British Columbia, Canada. *The Holocene* 23(11): 1572–83.

Mathewes, R. W., Lian, O. B., Clague, J. J., and Huntley, M. J. (2015). Early Wisconsinan (MIS 4) glaciation on Haida Gwaii, British Columbia, and implications for biological refugia. *Canadian Journal of Earth Sciences* 52(11): 939–51.

Miller, J. (2000). *Tsimshian Culture: A Light Through the Ages.* University of Nebraska Press.

Nichols, D. (2020). Rationale for Allowable Annual Cut (AAC) Determinations for Tree Farm Licence 58, Tree Farm Licence 60, and Timber Supply Area 25. British Columbia Ministry of Forests, Lands and Natural Resource Operations and Rural Development. https://www2.gov.bc.ca/assets/gov/farming-natural-resources-and-industry/forestry/stewardship/forest-analysis-inventory/tsr-annual-allowable-cut/hg_mu_ra_2020.pdf.

Ogilvie, R. T. (1994). *Rare and Endemic Vascular Plants of Gwaii Haanas (South Moresby) Park, Queen Charlotte Islands, British Columbia.* Forest Resource Development Agreement Report. No. 214. Royal British Columbia Museum.

Pojar, J. (2008). Changes in vegetation of Haida Gwaii in historical time. In *Lessons from the Islands: Introduced Species and What They Tell Us About How Ecosystems Work (Special Publication)*, edited by A. J. Gaston, T. E. Golumbia, J.-L. Martin, and S. T. Sharpe. Canadian Wildlife Service, Environment Canada, 32–36.

Ryan, T. (S.) L. (2014). *Territorial jurisdiction: The cultural and economic significance of eulachon* Thaleichthys pacificus *in the north-central coast region of British Columbia.* PhD dissertation, University of British Columbia. http://hdl.handle.net/2429/46515.

Simard, S. W., and Ryan, T. (S.) L. (2020). Feedback on the Haida Gwaii Timber Supply Review. The Mother Tree Project. https://mothertreeproject.org/2020/01/14/feedback-haida-gwaii-tsr/.

Takeda, L. (2014). *Islands' Spirit Rising: Reclaiming the Forests of Haida Gwaii.* UBC Press.

Wiggins, G. G., ed. (1996). *Proceedings of the Cedar Symposium: Growing Western Redcedar and Yellow-Cypress on the Queen Charlotte Islands/Haida Gwaii.* Canada-British Columbia South Moresby Forest Replacement Account.

Williams-Davidson, T. L. (2022). *Ts' uu JaasGalang hlGaajuu: Cedar sisters framework.* Doctoral dissertation, University of British Columbia.

8: CARBON BOMB

Abdenur, A. E. (2022). *The Glasgow Leaders' Declaration on Forests: Déjà Vu or Solid Restart?* United Nations University. New York. https://collections.unu.edu/view/UNU:8669.

Achat, D., Fortin, M., Landmann, G., et al. (2015). Forest soil carbon is threatened by intensive biomass harvesting. *Nature Scientific Reports* 5: 15991.

Beese, W. J., Deal, J., Dunsworth, B. G., et al. (2019). Two decades of variable retention in British Columbia: A review of its implementation and effectiveness for biodiversity conservation. *Ecological Processes* 8(1): 1–22.

Beillouin, D., Corbeels, M., Demenois, J., et al. (2023). A global meta-analysis of soil organic carbon in the Anthropocene. *Nature Communications* 14(1): 3700.

Berger, A. L., Palik, B., D'Amato, A. W., et al. (2013). Ecological impacts of energy-wood harvests: Lessons from whole-tree harvesting and natural disturbance. *Journal of Forestry* 111(2): 139–53.

Betts, M. G., Wolf, C., Ripple, W. J., et al. (2017). Global forest loss disproportionately erodes biodiversity in intact landscapes. *Nature* 547(7664): 441–44.

Betts, M. G., Yang, Z., Gunn, J. S., and Healey, S. P. (2024). Congruent long-term declines in carbon and biodiversity are a signature of forest degradation. *Global Change Biology* 30: e17541.

British Columbia Ministry of Forests Lands and Natural Resource Operations. (2017). Harvest data from Vegetation Resources Inventories (VRI) and Consolidated Cutblocks. BC Data Catalogue Metadata available at: https://catalogue.data.gov.bc.ca/dataset/b1b647a6-f271-42e0-9cd0-89ec24bce9f7.

Buotte, P. C., Law, B. E., Ripple, W. J., and Berner, L. T. (2019). Carbon sequestration and biodiversity co-benefits of preserving forests in the western United States. *Ecological Applications* 30: e02039.

Buotte, P. C., Levis, S., Law, B. E., et al. (2019). Near-future forest vulnerability to drought and fire varies across the western United States. *Global Change Biology* 25: 290–303.

Cambi, M., Certini, G., Neri, F., and Marchi, E. (2015). The impact of heavy traffic on forest soils: A review. *Forest Ecology and Management* 338: 124–38.

Canadian Forest Inventory Committee and Canadian Forest Service. (2008). *Canada's National Forest Inventory: Ground Sampling Guidelines.* Version 5.0. www.worldcat.org/oclc/606580128.

Crowther, T. W., Todd-Brown, K. E., Rowe, C. W., et al. (2016). Quantifying global soil carbon losses in response to warming. *Nature* 540(7631): 104–8.

Food and Agriculture Organization of the United Nations. (2020). *Global Forest Resources Assessment 2020—Key findings.* https://openknowledge.fao.org/items/d6f0df61-cb5d-4030-8814-0e466176d9a1.

Food and Agriculture Organization of the United Nations. (2024). *The State of the World's Forests 2024—Forest-Sector Innovations Towards a More Sustainable Future.* https://doi.org/10.4060/cd1211en.

Hagerman, S. M., Dowlatabadi, H., and Satterfield, T. (2010). Observations on drivers and dynamics of environmental policy change: Insights from 150 years of forest management in British Columbia. *Ecology and Society* 15(1).

Hagerman, S. M., Jones, M. D., Bradfield, G. E., et al. (1999). Effects of clear-cut logging on the diversity and persistence of ectomycorrhizae at a subalpine forest. *Canadian Journal of Forest Research* 29: 124–34.

Harmon, M. E., Moreno, A., and Domingo, J. B. (2009). Effects of partial harvest on the carbon stores in Douglas-fir/western hemlock forests: A simulation study. *Ecosystems* 12: 777–91. https://doi.org/10.1007/s10021-009-9256-2.

Hendrickson, O. Q., Chatarpaul, L., and Burgess, D. (1989). Nutrient cycling following whole-tree and conventional harvest in northern mixed forest. *Canadian Journal of Forest Research* 19(6): 725–35.

International Panel on Climate Change (2014): *Climate Change 2014: Synthesis Report. Contribution of Working Groups I, II and III to the Fifth Assessment Report of the Intergovernmental Panel on Climate Change.* Core writing team, R. K. Pachauri and L. A. Meyer. IPCC. Geneva, Switzerland.

International Panel on Climate Change. (2019). *Special Report on Climate Change and Land.* August. IPCC. https://www.ipcc/ch/srccl/.

International Panel on Climate Change. (2022). *Climate Change 2022: Impacts, Adaptation, and Vulnerability. Contribution of Working Group II to the Sixth Assessment Report of the Intergovernmental Panel on Climate Change.* Edited by H.-O. Pörtner, D. C. Roberts, M. Tignor, et al. Cambridge University Press. https://doi.org/10.1017/9781009325844.

James, J., and Harrison, R. (2016). The effect of harvest on forest soil carbon: A meta-analysis. *Forests* 7: 308.

Johnson, D. W., and Curtis, P. S. (2001). Effects of forest management on soil C and N storage: Meta-analysis. *Forest Ecology and Management* 140: 227–38.

Kimmins, J. P. (1976). Evaluation of the consequences for future tree productivity of the loss of nutrients in whole-tree harvesting. *Forest Ecology and Management* 1: 169–83.

Lake, H. (2024). *Motivations for alternative silviculture.* Master's thesis, Transatlantic Forestry Master's Program, Joensuu and University of British Columbia.

Law, B. E., Hudiburg, T. W., Berner, L. T., et al. (2018). Land use strategies to mitigate climate change in carbon dense temperate forests. *Proceedings of the National Academy of Sciences of the United States of America* 115: 3663–68.

Law, B. E., Moomaw, W. R., Hudiburg, T. W., et al. (2022). Creating strategic reserves to protect forest carbon and reduce biodiversity losses in the United States. *Land* 11(5): 721.

McKibben, B. (2022). In a world on fire, stop burning things. *The New Yorker.* https://www.newyorker.com/news/essay/in-a-world-on-fire-stop-burning-things.

Mo, L., Zohner, C. M., Reich, P. B., et al. (2023). Integrated global assessment of the natural forest carbon potential. *Nature* 624(7990): 92–101.

Mologni, O., Lahrsen, S., and Roeser, D. (2024). Automated production time analysis using FPDat II onboard computers: A validation study based on whole-tree ground-based harvesting operations. *Computers and Electronics in Agriculture* 222: 109047.

Nave, L. E., Vance, E. D., Swanston, C. W., and Curtis, P. S. (2010). Harvest impacts on soil carbon storage in temperate forests. *Forest Ecology and Management* 259(5): 857–66.

Pearse, P. H. (1990). *Introduction to Forestry Economics.* UBC Press.

Puettmann, K. J., Wilson, S. M., Baker, S. C., et al. (2015). Silvicultural alternatives to conventional even-aged forest management: What limits global adoption? *Forest Ecosystems* 2: 1–16.

Rähn, E., Tedersoo, L., Adamson, K., et al. (2023). Rapid shift of soil fungal community compositions after clear-cutting in hemiboreal coniferous forests. *Forest Ecology and Management* 544: 121211.

Roach, W. J., Simard, S. W., and Snyder, E. N. (2024). Downed woody debris varies with climate and harvesting treatment in Douglas-fir forests of British Columbia, Canada. *Frontiers in Forests and Global Change* 7: 1397142.

Robinson, A. J., Defrenne, C. E., Roach, W. J., et al. (2022). Harvesting intensity and aridity are more important than climate change in affecting future carbon stocks of Douglas-fir forests. *Frontiers in Forests and Global Change* 5: 934067.

Samuelson, P. A. (2018). Economics of Forestry in an Evolving Society. In *Classic Papers in Natural Resource Economics Revisited.* Routledge, 355–78.

Sands, B., Machado, M. R., White, A., et al. (2023). Moving towards an anti-colonial definition for regenerative agriculture. *Agriculture and Human Value* 40: 1697–1716.

Sasaki, T., Konno, M., Hasegawa, Y., et al. (2019). Role of mycorrhizal associations in tree spatial distribution patterns based on size class in an old-growth forest. *Oecologia* 189, 971–80.

Simard, S. W., Durall, D. M., and Jones, M. D. (1997). Carbon allocation and carbon transfer between *Betula papyrifera* and *Pseudotsuga menziesii* seedlings using a 13C pulse-labeling method. *Plant and Soil* 191: 41–55.

Simard, S. W., Roach, W. J., Defrenne, C. E., et al. (2020). Harvest intensity effects on carbon stocks and biodiversity are dependent on regional climate in Douglas-fir forests of British Columbia. *Frontiers in Forests and Global Change* 3: 88.

Smith, C. T., Briggs, R. D., Stupak, I., et al. (2022). Effects of whole-tree and stem-only clearcutting on forest floor and soil carbon and nutrients in a balsam fir (*Abies balsamea* (L.) Mill.) and red spruce (*Picea rubens* Sarg.) dominated ecosystem. *Forest Ecology and Management* 519: 120325.

Summers, B. (2024). *The interactive effects of harvesting intensity and climatic aridity on understory plant communities and Douglas-fir seedlings.* Master of science thesis, University of British Columbia. https://open.library.ubc.ca/soa/cIRcle/collections/ubctheses/24/items/1.0444028.

Tahvonen, O. (2016). Economics of rotation and thinning revisited: The optimality of clearcuts versus continuous cover forestry. *Forest Policy and Economics* 62: 88–94.

Tedersoo, L., Bahram, M., and Zobel, M. (2020). How mycorrhizal associations drive plant population and community biology. *Science* 367: eaba1223.

Tikina, A. V. (2008). Forest certification: Are we there yet? *CABI Reviews*. https://doi.org/10.1079/PAVSNNR20083018.

Villalobos, L., Coria, J., and Nordén, A. (2018). Has forest certification reduced forest degradation in Sweden? *Land Economics* 94(2): 220–38.

Walmsley, J. D., Jones, D. L., Reynolds, B., et al. (2009). Whole tree harvesting can reduce second rotation forest productivity. *Forest Ecology and Management* 257(3): 1104–11.

Zhou, D., Zhao, S. Q., Liu, S., and Oeding, J. (2013). A meta-analysis on the impacts of partial cutting on forest structure and carbon storage. *Biogeosciences* 10: 3691–703.

9: PLACE MATTERS

Boan, J., and Plotkin, R. (2025). *Counting on Canada's Commitments: To Halt and Reverse Forest Degradation by 2030, Canada Must First Admit It Has a Problem.* Forest-Degradation-in-Canada-R-25-04-A_06-1.pdf.

Brackett, A. E., Still, C. J., and Puettmann, K. J. (2024). Residual canopy cover provides buffering of near-surface temperatures, but benefits are limited under extreme conditions. *Canadian Journal of Forest Research* 54(9): 1018–1031.

Constantinou, A., Burton, A. C., Simard, S. W., and Hodder, D. (2025). Initial effects of clearcutting and partial retention forest harvesting methods on some small mammals in northern British Columbia. *Journal of Ecosystems and Management* 25(1): 12—pp.

Constantinou, A., Burton, A. C., Simard, S. W., and Hodder, D. (2025). Responses of large mammals to forest harvesting treatments across a climate gradient. *Journal of Ecosystems and Management,* submitted.

De Frenne, P., Zellweger, F., Rodríguez-Sánchez, F., et al. (2019). Global buffering of temperatures under forest canopies. *Nature Ecology and Evolution* 3(5): 744–49.

Goward, T., Coxson, D., and Gauslaa, Y. (2024). The Manna Effect—A review of factors influencing hair lichen abundance for Canada's endangered Deep-Snow Mountain Caribou (*Rangifer arcticus montanus*). *The Lichenologist* 56(4): 1–15.

Harris, T. C., Roach, W. J., Miller, E. M., and Simard, S. W. (2025). The interactive role of climatic transfer distance and overstory retention on Douglas-fir seedling survival and height growth in interior British Columbia. *Global Change Biology* 31(1): e70027.

Harris, T. C., Roach, W. J., Miller, E. M., and Simard, S. W. (2025). The interactive role of regional climate and overstory retention on planted seedling performance in interior British Columbia. *Global Change Biology*, submitted.

Owen, Brenna. (2025). Old-growth logging was "goal" of Interfor: B.C. Forest Appeals Commission decision. *The Canadian Press*. https://www.ctvnews.ca/vancouver/article/old-growth-logging-was-goal-of-interfor-bc-forest-appeals-commission-decision/.

Serrouya, R., and Robert, G. D. (2008). The influence of forest cover on mule deer habitat selection, diet, and nutrition during winter in a deep-snow ecosystem. *Forest Ecology and Management* 256(3): 452–61.

Simard, S. W., Roach, W. J., Beauregard, J., et al. (2021). Partial retention of legacy trees protects mycorrhizal inoculum potential, biodiversity, and soil resources while promoting natural regeneration of interior Douglas-fir. *Frontiers in Forests and Global Change* 3: 620436.

Simard, S. W., Roach, W. J., Defrenne, C. E., et al. (2020). Harvest intensity effects on carbon stocks and biodiversity are dependent on regional climate in Douglas-fir forests of British Columbia. *Frontiers in Forests and Global Change* 3: 88.

Smith, K. G., Ficht, E. J., Hobson, D., et al. (2000). Winter distribution of woodland caribou in relation to clear-cut logging in west-central Alberta. *Canadian Journal of Zoology* 78(8): 1433–40.

Suggitt, A. J., Gillingham, P. K., Hill, J. K., et al. (2011). Habitat microclimates drive fine-scale variation in extreme temperatures. *Oikos* 120(1): 1–8.

Summers, B. (2024). *The interactive effects of harvesting intensity and climatic aridity on understory plant communities and Douglas-fir seedlings*. Master of science thesis, University of British Columbia.

Summers, B., Law, D., Roach, W. J., et al. (2025). Partial retention of overstory trees increasingly protects understory plant communities as aridity deepens in interior Douglas-fir forests of British Columbia. *Global Change Biology*, submitted.

Terry, E. L., McLellan, B. N., and Watts, G. S. (2000). Winter habitat ecology of mountain caribou in relation to forest management. *Journal of Applied Ecology* 37(4): 589–602.

Teskey, R., Wertin, T., Bauweraerts, I., et al. (2015). Responses of tree species to heat waves and extreme heat events. *Plant, Cell and Environment* 38(9): 1699–712.

Waterhouse, M. J., Armleder, H. M., and Nemec, A. F. (2011). Terrestrial lichen response to partial cutting in lodgepole pine forests on caribou winter range in west-central British Columbia. *Rangifer* 119–34.

10: RECIPROCITY

Alexander, M. E. (2010). Surface fire spread potential in trembling aspen during summer in the Boreal Forest Region of Canada. *The Forestry Chronicle* 86(2): 200–12.

Amyntas, A., Berti, E., Gauzens, B., et al. (2023). Niche complementarity among plants and animals can alter the biodiversity-ecosystem functioning relationship. *Functional Ecology* 37(10): 2652–65.

Armstrong, C. G., Grenz, J., Zyp-Loring, J., et al. (2024). Ethnoecological perspectives on environmental stewardship: Tenets and basis of reciprocity in Gitxsan and nłeʔkepmx (Nlaka'pamux) Territories. *People and Nature* 1–13.

Arnold, C. (2021). Reciprocal relationships with trees: Rekindling indigenous wellbeing and identity through the Yuin ontology of oneness. *Australian Geographer* 52: 131–47.

Arrow Lakes News. (2021). Old-growth logging protest attracts crowd at B.C. Forest Minister's Castlegar office. https://www.arrowlakesnews.com/news/old-growth-logging-protest-attracts-crowd-at-b-c-forest-ministers-castlegar-office-4611579.

Bennett, J. N., and Prescott, C. E. (2004). Organic and inorganic nitrogen nutrition of western red cedar, western hemlock and salal in mineral N-limited cedar-hemlock forests. *Oecologia* 141: 468–76.

Brooks, J. R., Meinzer, F. C., Warren, J. M., et al. (2006). Hydraulic redistribution in a Douglas-fir forest: Lessons from system manipulations. *Plant, Cell and Environment* 29(1): 138–50.

Canadian Broadcasting Corporation. (2016). Stolen Children: Residential School survivors speak out. CBC News. (YouTube) https://www.youtube.com/watch?v=vdR9HcmiXLA.

Canadian Broadcasting Corporation. (2021). A look back at the 2021 B.C. wildfire season. CBC News. https://www.cbc.ca/news/canada/british-columbia/bc-wildfires-2021-timeline-1.6197751.

Canadian Broadcasting Corporation. (2022). Coroner's report on B.C. heat-dome deaths calls for greater support for populations at risk. CBC News. https://www.cbc.ca/news/canada/british-columbia/bc-heat-dome-coroners-report-1.6480026.

Chiolerio, A., Gagliano, M., Pilia, S., et al. (2025). Bioelectrical synchronization of *Picea abies* during a solar eclipse. *Royal Society Open Science* 12(4): 241786.

Filotas, E., Parrott, L., Burton, P. J., et al. (2014). Viewing forests through the lens of complex systems science. *Ecosphere* 5(1): 1.

Gorzelak, M., Pickles, B. J., Asay, A. K., and Simard, S. W. (2015). Inter-plant communication through mycorrhizal networks mediates complex adaptive behaviour in plant communities. *Annals of Botany Plants* 7: plv050.

Grenz, J. (2024). *Medicine Wheel for the Planet: A Journey Toward Personal and Ecological Healing.* Knopf Canada.

Harrap, L. (2021). Letter: Save Box Mountain in Nakusp from logging. *Arrow Lakes News.* https://www.arrowlakesnews.com/opinion/letter-save-box-mountain-in-nakusp-from-logging-4612343.

Jabr, Ferris. (2020). The Social Life of Forests. *The New York Times.* https://www.nytimes.com/interactive/2020/12/02/magazine/tree-communication-mycorrhiza.html.

Jain, P., Sharma, A. R., Acuna, D. C., et al. (2024). Record-breaking fire weather in North America in 2021 was initiated by the Pacific Northwest heat dome. *Communications Earth and Environment* 5: 202.

Kimmerer, R. W. (2011). Restoration and reciprocity: The contributions of traditional ecological knowledge. In *Human Dimensions of Ecological Restoration: Integrating Science, Nature, and Culture.* 257–76. Washington, DC: Island Press/Center for Resource Economics. doi:10.5822/978-1-61091-039-2_18.

Kimmerer, R. W. (2017). The Covenant of Reciprocity. In *The Wiley Blackwell Companion to Religion and Ecology,* 368–81. Edited by J. Hart. Wiley Online Library.

Kimmerer, R. W. (2025). *The Serviceberry: Abundance and Reciprocity in the Natural World.* Scribner.

Kobylinski, A., and Fredeen, A. L. (2014). Vertical distribution and nitrogen content of epiphytic macrolichen functional groups in sub-boreal forests of central British Columbia. *Forest Ecology and Management* 329: 118–28.

Kranabetter, J. M., Dawson, C. R., and Dunn, D. E. (2007). Indices of dissolved organic nitrogen, ammonium and nitrate across productivity gradients of boreal forests. *Soil Biology and Biochemistry* 39(12): 3147–58.

Levin, S. A. (2005). Self-organization and the emergence of complexity in ecological systems. *BioScience* 55(12): 1075–79.

Metcalfe, B. (2021). 2021 Year in Review: Climate change hits home in Nelson. *Nelson Star.* https://www.nelsonstar.com/news/2021-year-in-review-climate-change-hits-home-in-nelson-4899032.

Oliveira, A. S., Silva, J. S., Guiomar, N., et al. (2023). The effect of broadleaf forests in wildfire mitigation in the WUI—A simulation study. *International Journal of Disaster Risk Reduction* 93: 103788.

Oliver, T. H., Heard, M. S., Isaac, N. J., et al. (2015). Biodiversity and resilience of ecosystem functions. *Trends in Ecology and Evolution* 30(11): 673–84.

Oliver, T. H., Isaac, N. J., August, T. A., et al. (2015). Declining resilience of ecosystem functions under biodiversity loss. *Nature Communications* 6(1): 10122.

Orrego, G. (2018). *Western hemlock regeneration on coarse woody debris is facilitated by linkage into a mycorrhizal network in an old-growth forest.* Master's thesis, University of British Columbia. https://doi.org/10.14288/1.0365590.

Perry, D. A. (1995). Self-organizing systems across scales. *Trends in Ecology and Evolution* 10(6): 241–44.

Perry, D. A., Oren, R., and Hart, S. C. (2008). *Forest Ecosystems.* Johns Hopkins University Press.

Pickles, B. J., Wilhelm, R., Asay, A. K., et al. (2017). Transfer of 13C between paired Douglas-fir seedlings reveals plant kinship effects and uptake of exudates by ectomycorrhizas. *New Phytologist* 214(1): 400–11.

Roach, W. J., Simard, S. W., Defrenne, C. E., et al. (2021). Tree diversity, site index, and carbon storage decrease with aridity in Douglas-fir forests in western Canada. *Frontiers in Forests Global Change* 4: 682076.

Sajedi, T., Prescott, C. E., Seely, B., and Lavkulich, L. M. (2012). Relationships among soil moisture, aeration and plant communities in natural and harvested coniferous forests in coastal British Columbia, Canada. *Journal of Ecology* 100(3): 605–18.

Schoonmaker, A. L., Teste, F. P., Simard, S. W., and Guy, R. D. (2007). Tree proximity, soil pathways and common mycorrhizal networks: Their influence on utilization of redistributed water by understory seedlings. *Oecologia* 154: 455–66.

Simard, S. W. (2009). The foundational role of mycorrhizal networks in self-organization of interior Douglas-fir forests. *Forest Ecology and Management* 258S: S95—S107.

Simard, S. W. (2018). Mycorrhizal networks facilitate tree communication, learning and memory. In *Memory and Learning in Plants,* edited by F. Baluska, M. Gagliano, and G. Witzany. Springer, 191–213.

Simard, S. W. (2021). *Finding the Mother Tree: Discovering the Wisdom of the Forest.* Penguin Random House.

Simard, S. W., Beiler, K. J., Bingham, M. A., et al. (2012). Mycorrhizal networks: Mechanisms, ecology and modelling. Invited review. *Fungal Biology Reviews* 26: 39–60.

Simard, S. W., Hagerman, S. M., Sachs, D. L., et al. (2005). Conifer growth, *Armillaria ostoyae* root disease and plant diversity responses to broadleaf competition reduction in temperate mixed forests of southern interior British Columbia. *Canadian Journal of Forest Research* 35: 843–59.

Simard, S. W., Jones, M. D., Durall, D. M., et al. (1997). Reciprocal transfer of carbon isotopes between ectomycorrhizal *Betula papyrifera* and *Pseudotsuga menziesii*. *New Phytologist* 137: 529–42.

Simard, S. W., Perry, D. A., Jones, M. D., et al. (1997). Net transfer of carbon between tree species with shared ectomycorrhizal fungi. *Nature* 388: 579–82.

Simard, S. W., Radosevich, S. R., Sachs, D. L., and Hagerman, S. M. (2006). Evidence for competition/facilitation trade-offs: Effects of Sitka alder density on pine regeneration and soil productivity. *Canadian Journal of Forest Research* 36: 1286–98.

Teste, F. P., Jones, M. D., and Dickie, I. A. (2020). Dual-mycorrhizal plants: Their ecology and relevance. *New Phytologist* 225(5): 1835–51.

Titus, B. D., Prescott, C. E., Maynard, D. G., et al. (2006). Post-harvest nitrogen cycling in clearcut and alternative silvicultural systems in a montane forest in coastal British Columbia. *The Forestry Chronicle* 82(6): 844–59.

Warner, E., Cook-Patton, S. C., Lewis, O. T., et al. (2023). Young mixed planted forests store more carbon than monocultures—A meta-analysis. *Frontiers in Forests Global Change* 6: 1226514.

Wilson, Jeremy. (1998). *Talk and Log: Wilderness Politics in British Columbia.* University of British Columbia Press.

11: THE FOREST DEFENDERS

Araujo, E. C. G., Sanquetta, C. R., Dalla Corte, A. P., et al. (2023). Global review and state-of-the-art of biomass and carbon stock in the Amazon. *Journal of Environmental Management* 331: 117251.

Arbess, Bobby. (2020). Fairy Creek Blockade. Focus on Victoria. https://www.focusonvictoria.ca/forests/32/.

Artelle, K. A., Zurba, M., Bhattacharyya, J., et al. (2019). Supporting resurgent Indigenous-led governance: A nascent mechanism for just and effective conservation. *Biological Conservation* 240: 108284.

Austen, Ian. (2021). British Columbia Judge Ends Protest Injunction at Fairy Creek. *The New York Times.* https://www.nytimes.com/2021/10/01/world/canada/british-columbia-fairy-creek-protests.html.

The Canadian Press. (2025). B.C. extends deferral of logging in Fairy Creek amid reports of tree spiking. CBC News. https://www.cbc.ca/news/canada/british-columbia/fairy-creek-defer-old-growth-logging-extension-2025-1.7445346.

Churchland, C., Bengtson, P., Prescott, C. E., and Grayston, S. J. (2021). Dispersed variable-retention harvesting mitigates N losses on harvested sites in conjunction with changes in soil microbial community structure. *Frontiers in Forests and Global Change* 3: 609216.

Committee on the Status of Endangered Wildlife in Canada. (2010). *COSEWIC Assessment and Status Report on the Oldgrowth Specklebelly* Pseudocyphellaria rainierensis *in Canada.* COSEWIC. http://www.publications.gc.ca/site/eng/9.567433/marcXml.html?MODS=1.

Cox, Sarah. (2021). Inside the Pacheedaht Nation's stand on Fairy Creek logging blockades. *The Narwhal.* https://thenarwhal.ca/pacheedaht-fairy-creek-bc-logging/.

Davis, H., (1996). *Characteristics and selection of winter dens by black bears in coastal British Columbia*. Bachelor of science thesis, Simon Fraser University. https://api.semantic scholar.org/CorpusID:130896810.

Davis, H., Hamilton, A. N., Harestad, A. S., and Weir, R. D. (2012). Longevity and reuse of black bear dens in managed forests of coastal British Columbia. *Journal of Wildlife Management* 76(3): 523–27.

DellaSala, D. A., Keith, H., Sheehan, T., et al. (2022). Estimating carbon stocks and stock changes in Interior Wetbelt forests of British Columbia, Canada. *Ecosphere* 13(4): e4020.

Ditidaht, Huu-ay-aht, and Pacheedaht First Nations. Hišuk ma c̓awak Declaration, huuayaht.org/wp-content/uploads/2021/06/declaration-FINAL-signedpdf.pdf.

Ecojustice. (2024). Protecting B.C.'s old-growth on International Day of Forests. https:// ecojustice.ca/news/international-day-of-forests-are-b-c-s-forests-for-the-people-or -profits/.

Franklin, J. F., Cromack, K., Jr., Denison, W., et al. (1981). Ecological characteristics of old-growth Douglas-fir forests. Gen. Tech. Rep. PNW-118. USDA Forest Service, Pacific Northwest Forest and Range Experiment Station.

Franklin, J. F., Shugart, H. H., and Harmon, M. E. (1987). Tree death as an ecological process. *BioScience* 37(8): 550–56.

Gorley, A., and Merkel, G. (2020). A new future for old forests: A strategic review of how British Columbia manages for old forests within its ancient ecosystems. British Columbia Ministry of Forests, Lands, Natural Resources Operations, and Rural Development. https://www2.gov.bc.ca/assets/gov/farming-natural-resources-and -industry/forestry/stewardship/old-growth-forests/strategic-review-20200430.pdf.

Guichon, S. H. A. (2015). *Mycorrhizal fungi: Unlocking their ecology and role in the establishment and growth performance of different conifer species in nutrient-poor coastal forests*. Doctoral dissertation, University of British Columbia. https://open.library .ubc.ca/soa/cIRcle/collections/ubctheses/24/items/1.0221252.

Hunter, Justine. (2021). How an American teen's pandemic lockdown launched B.C.'s biggest logging protest in decades. *The Globe and Mail*. https://www.theglobeand mail.com/canada/british-columbia/article-how-an-american-teens-pandemic-lockd own-launched-bcs-biggest-logging/.

iNaturalist. Fairy Creek Research. https://www.inaturalist.org/projects/fairy-creek -research.

Ingham, R. E., Trofymow, J. A., Ingham, E. R., and Coleman, D. C. (1985). Interactions of bacteria, fungi, and their nematode grazers: Effects on nutrient cycling and plant growth. *Ecological Monographs* 55: 119–40.

Joseph, B. (2018). *21 Things You May Not Know About the Indian Act: Helping Canadians Make Reconciliation with Indigenous Peoples a Reality*. Indigenous Relations Press.

Kulkary, Akshay. (2025). Green Party deputy leader gets jail time for Fairy Creek protests. CBC News. https://www.cbc.ca/news/canada/british-columbia/angela-davidson -rainbow-eyes-fairy-creek-1.7164288.

Labbé, Stefan. (2025). BC old-growth forests: Protection benefits valued at $10.9B. *Business in Vancouver*. https://www.biv.com/news/protecting-bc-old-growth-forests -could-yield-109b-in-benefits-report-finds-10481195.

Legree, D. (2023). A watershed moment at the Fairy Creek watershed? BC forestry policy and punctuated equilibrium theory. https://qspace.library.queensu.ca/server/api /core/bitstreams/214628ee-6666-4457-8b43-c23a3c2b00ec/content.

Lindsay, Evan. (2025). Deputy Green leader Rainbow Eyes sees jail time slashed in old growth protest. *Terrace Standard.* https://www.terracestandard.com/home2/deputy-green-leader-rainbow-eyes-sees-jail-time-slashed-in-old-growth-protest-7937362.
Luyssaert, S., Schulze, E.-D., Börner, A., et al. (2008). Old-growth forests as global carbon sinks. *Nature* 455: 213–15.
Maingon, Loys. (2021). Fairy Creek Species at Risk. *Watershed Sentinel.* https://watershedsentinel.ca/article/fairy-creek-species-at-risk/.
Mapes, Lynda V. (2021). B.C. government continues logging of old growth as 2-year protest in the woods drags on. *The Seattle Times.* https://www.seattletimes.com/seattle-news/environment/b-c-government-continues-logging-of-old-growth-as-2-year-protest-in-the-woods-drags-on/.
Mapes, Lynda V. (2025). *The Trees Are Speaking: Dispatches from the Salmon Forests.* University of Washington Press.
Miller, S. L., Raphael, M. G., Falxa, G. A., et al. (2012). Recent population decline of the Marbled Murrelet in the Pacific Northwest. *The Condor* 114(4): 771–81.
Morton, C., Beukema, S., Poulsen, F., et al. (2025). *The Economic Value of Old-growth Forests in BC: Analysis of Old-growth Management Scenarios in Two Timber Supply Areas.* ESSA Technologies Ltd., 20.
Morton, C., Trenholm, R., Beukema, S., et al. (2021). *Economic valuation of old growth forests on Vancouver Island: Pilot study; Phase 2—Port Renfrew Pilot.* ESSA Technologies Ltd. for the Ancient Forest Alliance. ancientforestalliance.org/old-growth-economic-report.
The Narwhal. (2021). Fairy Creek Blockade and Old-Growth Forests in B.C. https://thenarwhal.ca/topics/fairy-creek-blockade/.
Neilson, J., Maingon, L., and Lavdovsky, N. (2022). Without an over-arching biodiversity protection act, what protections exist for biodiversity in British Columbia? A case study of Oldgrowth Specklebelly Lichen (*Pseudocyphellaria rainierensis*). *The Canadian Field-Naturalist* 136(2): 192–96.
Old Growth Forest Ecology. https://oldgrowthforestecology.org/.
Piatt, J. F., Kuletz, K. J., Burger, A. E., et al. (2007). Status review of the marbled murrelet (*Brachyramphus marmoratus*) in Alaska and British Columbia. Open-file report 2006-1387. U.S. Department of the Interior, U.S. Geological Service.
Pierce, Daniel. (2018). 25 Years after the War in the Woods: Why B.C.'s forests are still in crisis. *The Narwhal.* https://thenarwhal.ca/25-years-after-clayoquot-sound-blockades-the-war-in-the-woods-never-ended-and-its-heating-back-up/.
Pregitzer, K. S., and Euskirchen, E. S. (2004). Carbon cycling and storage in world forests: Biome patterns related to forest age. *Global Change Biology* 10: 2052–77.
Prescott, C. E., and Grayston, S. J. (2023). TAMM review: Continuous root forestry—Living roots sustain the belowground ecosystem and soil carbon in managed forests. *Forest Ecology and Management* 532: 120848.
Price, K., Daust, D., Daust, K., and Holt, R. (2023). Estimating the amount of British Columbia's "big-treed" old growth: Navigating messy indicators. *Frontiers in Forests and Global Change* 5: 958719.
Price, K., Holt, R., and Daust, D. (2020). BC's Old Growth Forests: A Last Stand for Biodiversity. Veridian Ecological. https://veridianecological.ca/wp-content/uploads/2020/05/bcs-old-growth-forest-report-web.pdf.
Price, K., Holt, R. F., and Daust, D. (2021). Conflicting portrayals of remaining old growth: The British Columbia case. *Canadian Journal of Forest Research* 51(5): 742–52.
Province of British Columbia. Old growth deferral areas. https://www2.gov.bc.ca/gov

/content/industry/forestry/managing-our-forest-resources/old-growth-forests/deferral-areas.

Province of British Columbia. TFL 46. https://www2.gov.bc.ca/gov/content/industry/forestry/forest-tenures/timber-harvesting-rights/tfl/tfl-46.

Richard, Graham (Jaahljuu). (2025). Haida Gwaii court ruling against forestry giant is a win for Indigenous Rights. Court rejects logging company's claim to conservation-related losses. *The Narwhal.* https://thenarwhal.ca/haida-gwaii-management-council-teal-jones-reconciliation-losses/.

Roach, W. J., Simard, S. W., Defrenne, C. E., et al. (2021). Tree diversity, site index, and carbon storage decrease with aridity in Douglas-fir forests in western Canada. *Frontiers in Forests and Global Change* 4: 682076.

Robinson, A. J., Defrenne, C. E., Roach, W. J., et al. (2022). Harvesting intensity and aridity are more important than climate change in affecting future carbon stocks of Douglas-fir forests. *Frontiers in Forests and Global Change* 5: 934067.

Sierra Club of British Columbia. (2019). Poll shows nine in ten British Columbians support action to protect endangered old-growth forest. https://sierraclub.bc.ca/forestpoll/.

Simard, S. W., Roach, W. J., Defrenne, C. E., et al. (2020). Harvest intensity effects on carbon stocks and biodiversity are dependent on regional climate in Douglas-fir forests of British Columbia. *Frontiers in Forests and Global Change* 3: 88.

Simmons, Matthew. (2021). Ecologist Suzanne Simard offers solutions to B.C.'s forest woes. *The Narwhal.* https://thenarwhal.ca/suzanne-simard-mother-tree-profile/.

Smithwick, E. A. H., Harmon, M. E., Remillard, S. M., et al. (2002). Potential upper bounds of carbon stores in forests of the Pacific Northwest. *Ecological Applications* 12(5): 1303–17.

Species at Risk Act Registry. (2021). Species summary: Oldgrowth Specklebelly Lichen (*Pseudocyphellaria rainierensis*). Government of Canada. https://species-registry.canada.ca/index-en.html#/species/126-423.

Species at Risk Act Registry. (2021). Species summary: Western Screech-owl kennicotti subspecies (*Megascops kenicotti kennicotti*). Government of Canada. https://species-registry.canada.ca/index-en.html#/ species/719-101.

Stand.Earth. (2021). Land Defenders and old growth forests at high risk as Ada'itsx (Fairy Creek) becomes the largest act of civil disobedience in Canadian history. https://stand.earth/press-releases/land-defenders-and-old-growth-forests-at-high-risk-as-adaitsx-fairy-creek-becomes-the-largest-act-of-civil-disobedience-in-canadian-history/.

Tafler, Sid. (2021). Old-Growth Defenders Have a Formidable Ally in Suzanne Simard. *The Tyee.* https://thetyee.ca/News/2021/08/11/Old-Growth-Defenders-Formidable-Ally-Suzanne-Simard/.

Trofymow, J. A., and Blackwell, B. A. (1998). Changes in ecosystem mass and carbon distributions in coastal forest chronosequences. *Northwest Science* 72: 40–42.

Trofymow, J. A., Stinson, G., and Kurz, W. A. (2008). Derivation of a spatially explicit 86-year retrospective carbon budget for a landscape undergoing conversion from old-growth to managed forests on Vancouver Island, BC. *Forest Ecology and Management* 256(10): 1677–91.

Wardle, D. A. (1999). How soil foodwebs make plants grow. *Trends in Ecology and Evolution* 14 (11): 418–20.

Waters, Shannon. (2025). B.C. government aims to permanently protect Fairy Creek. *The Narwhal.* https://thenarwhal.ca/bc-fairy-creek-protection-pledge/.

Watt, T. J. (2021). Fairy Creek Headwaters in the Snow. TJ Watt Photography.
Wyton, M. (2023). B.C. MP introduces motion to end old-growth logging on federal land, ban all exports by 2030. CBC News. https://www.cbc.ca/player/play/video/1.6834514.
Yunker, Zöe. (2021). Police Raid the Hub for Fairy Creek Blockaders. *The Tyee*. https://thetyee.ca/News/2021/08/10/Police-Raid-Fairy-Creek-Blockader-Hub/.

12: THE FIRST ECOLOGISTS

Armstrong, C. G., Clemente-Carvalho, R. B., Turner, N. J., et al. (2024). Genetic differentiation and precolonial Indigenous cultivation of hazelnut (*Corylus cornuta*, Betulaceae) in western North America. *Proceedings of the National Academy of Sciences* 121(48): e2402304121.
Armstrong, C. G., Lyons, N., McAlvay, A. C., et al. (2023). Historical ecology of forest garden management in L a xyuubm Ts' msyen and beyond. *Ecosystems and People* 19(1): 2160823.
Armstrong, C. G., Miller, J. E., McAlvay, A. C., et al. (2021). Historical indigenous land-use explains plant functional trait diversity. *Ecology and Society* 26(2): 6.
Asay, A. K., Simard, S. W., and Dudley, S. A. (2020). Altering neighborhood relatedness and species composition affects interior Douglas-fir size and morphological traits with context-dependent responses. *Frontiers in Ecology and Evolution* 8: 578524.
Awi'nakola. Awi'nakola—Tree of Life. https://www.awinakola.com.
Bliege Bird, R., McGuire, C., Bird, D. W., et al. (2020). Fire mosaics and habitat choice in nomadic foragers. *Proceedings of the National Academy of Sciences* 117(23): 12904–14.
Canadian Broadcasting Corporation. (2019). B.C. First Nation feeds hungry grizzlies 500 salmon carcasses. CBC News. https://www.cbc.ca/news/canada/british-columbia/mamalilikulla-first-nation-feeding-starving-grizzlies-1.5305820.
Glass, A. (2021). *Writing the Hamat̕sa: Ethnography, Colonialism and the Cannibal Dance*. University of British Columbia Press.
Hernandez, Jon. (2018). B.C. loggers aim to transition away from harvesting old growth—But it could take 90 years. CBC News. https://www.cbc.ca/news/canada/british-columbia/b-c-loggers-aim-to-transition-away-from-harvesting-old-growth-but-it-could-take-90-years-1.4735715.
Hoffman, K. M., Davis, E .L., Wickham, S. B., et al. (2021). Conservation of Earth's biodiversity is embedded in Indigenous fire stewardship. *Proceedings of the National Academy of Sciences* 118(32): e2105073118.
Kimmerer, R. (2011). Restoration and Reciprocity: The Contributions of Traditional Ecological Knowledge. In *Human Dimensions of Ecological Restoration*, edited by D. Egan, E. E. Hjerpe, and J. Abrams. Society for Ecological Restoration. Island Press. https://doi.org/10.5822/978-1-61091-039-2_18.
Kimmerer, R. W. (2013). *Braiding Sweetgrass: Indigenous Wisdom, Scientific Knowledge and the Teachings of Plants*. Milkweed Editions.
Kimmerer, R. W. and Lake, F. K. (2001). The role of indigenous burning in land management. *Journal of Forestry* 99(11): 36–41.
Kwakwaka'wakw Nation. Our Land, Our People, Living Tradition, The Kwakwaka'wakw Potlatch on the Northwest Coast. https://umistapotlatch.ca/notre_peuple-our_people-eng.php.
Le Guin, Ursula. (1972). *The Word for World Is Forest*. Doubleday.
Ma'amtagila Nation. We are the Ma'amtagila Nation. https://www.maamtagila.ca.

Mo, L., Zohner, C. M., Reich, P. B., et al. (2023). Integrated global assessment of the natural forest carbon potential. *Nature* 624: 92–101.
Nawalakw. https://nawalakw.com/.
Nelson, K'odi. K'odi Nelson—Executive Director—Nawalakw Healing Society. LinkedIn.
Richardson, Kelly. https://kellyrichardson.net.
Sierra Club of BC. (2021). Into the woods: An interview with coastal projects lead Mark Worthing. https://sierraclub.bc.ca/an-interview-with-coastal-projects-lead-mark-worthing/.
Simard, S. W. (2016). Nature's internet: How trees talk to each other in a healthy forest. TEDxSeattle. https://www.youtube.com/results?search_query=suzanne+simard+tedx.
Simard, S. W. (2016). Suzanne Simard: How trees talk to each other. TED Talk. https://www.ted.com/talks/suzanne_simard_how_trees_talk_to_each_other.
Smith, Stephanie. About—Independent Curators International. https://curatorsintl.org/about/collaborators/4391-stephanie-smith.
Talaga, Tanya. (2024). *The Knowing*. HarperCollins.
Tłalita'las. LAND Stewardship with Tłaḷita'las. Déjà Well. https://www.dejawell.com/workshops/land-stewardship-with-talitalas-karissa-glendale-of-traditional-kwakwakawakw-territory.
Turner, N. J. (2008). *The Earth's Blanket: Traditional Teachings for Sustainable Living*. D & M Publishers.
Vierros, M. (2017). Communities and blue carbon: The role of traditional management systems in providing benefits for carbon storage, biodiversity conservation and livelihoods. *Climatic Change* 140: 89–100. https://doi.org/10.1007/s10584-013-0920-3.
Walde, Paul. http://paulwalde.com/.
Walker, W., Baccini, A., Schwartzman, S., et al. (2014). Forest carbon in Amazonia: The unrecognized contribution of indigenous territories and protected natural areas. *Carbon Management* 5(5–6): 479–85.
Watershed Sentinel. (2023). Stop the Spray: Glyphosate, Forest Monoculture, and Fire. September 27, 2023. watershedsentinel.ca/articles/stop-the-spray/.

13: IN THE ARC OF A TURN

Asay, A. K. (2013). *Mycorrhizal facilitation of kin recognition in interior Douglas-fir (*Pseudotsuga menziesii *var.* glauca*)*. Master's of science thesis, University of British Columbia. https://open.library.ubc.ca/soa/cIRcle/collections/ubctheses/24/items/1.0103374.
Asay, A. K. (2019). *Influence of kin, density, soil inoculum potential and interspecific competition on interior Douglas-fir (*Pseudotsuga menziesii *var.* glauca*) performance and adaptive traits* (Doctoral dissertation, University of British Columbia). https://doi.org/10.14288/1.0387438.
Asay, A. K., Simard, S. W., and Dudley, S. A. (2020). Altering neighborhood relatedness and species composition affects interior Douglas-fir size and morphological traits with context-dependent responses. *Frontiers in Ecology and Evolution* 8: 578524.
DeLong, R., Lewis, K. J., Simard, S. W., and Gibson, S. (2002). Fluorescent pseudomonad population sizes baited from soils under pure birch, pure Douglas-fir and mixed forest stands and their antagonism toward *Armillaria ostoyae* in vitro. *Canadian Journal of Forest Research* 32: 2146–59.

District of Sicamous. (2022). Sicamous and Splatsin students explore Secwépemc pictographs. https://www.sicamous.ca/live-here/news/post/secwepemc-pictographs.

Harper, T. (2022). "A terrific person": Canadian baseball star remembered after tragic death at Whitewater Ski Resort. *Castlegar News.* https://www.castlegarnews.com/news/canada-baseball-star-amanda-asay-dies-after-accident-at-whitewater-ski-resort-4746787.

Karst, J., Jones, M. D., and Hoeksema, J. D. (2023). Positive citation bias and overinterpreted results lead to misinformation on common mycorrhizal networks in forests. *Nature Ecology and Evolution* 7(4): 501–11.

Sundstrom, S. M., and Allen, C. R. (2019). The adaptive cycle: More than a metaphor. *Ecological Complexity* 39: 100767.

14: TREES IN OUR BLOOD

Baleshta, K. E., Simard, S. W., and Roach, W. J. (2015). Effects of thinning paper birch on conifer productivity and understory plant diversity. *Scandinavian Journal of Forest Research* 30(8): 699–709.

Baumflek, M., Kassam, K. A., Ginger, C., and Emery, M. R. (2021). Incorporating biocultural approaches in forest management: Insights from a case study of Indigenous plant stewardship in Maine, USA and New Brunswick, Canada. *Society and Natural Resources* 34: 1155–73.

Blunt, Zoe. (2024). TimberWest Quietly Drops Herbicide Spray. *Watershed Sentinel.* https://watershedsentinel.ca/article/timberwest-drops-herbicide/.

Charnley, S., Fischer, A. P., and Jones, E. T. (2007). Integrating traditional and local ecological knowledge into forest biodiversity conservation in the Pacific Northwest. *Forest Ecology and Management* 246: 14–28.

Chief Adam Dick (Kwaxsistalla Wathl'thla), Daisy Sewid-Smith (Mayanilth), Kim Recalma-Clutesi (Oqwilowgwa), et al. (2022). "From the beginning of time": The colonial reconfiguration of native habitats and Indigenous resource practices on the British Columbia Coast. *FACETS* 7: 543–70. https://doi.org/10.1139/facets-2021-0092.

Cruikshank, Ainslee. (2024). A map of glyphosate use in B.C. forests. *The Narwhal.* https://thenarwhal.ca/bc-glyphosate-forestry-map/.

Cruz, O., Drouillard, B., Mund, M., and Pearse, B. (2018). Changing the Climate Around Estuarine Root Garden Management with the Kwakwaka'wakw. *In/Versions.* Changing-the-Climate-around-Estuarine-Root-Garden-Management-with-the-Kwakwakawakw.pdf.

Daily, G. (1995). Restoring value to the world's degraded lands. *Science* 269(5222): 350–54.

Daniels, L. D., Dickson-Hoyle, S., Baron, J. N., et al. (2024). The 2023 wildfires in British Columbia, Canada: Impacts, drivers, and transformations to coexist with wildfire. *Canadian Journal of Forest Research* 55: 1–18.

Dawe, Charlotte. (2023). Toxic glyphosate to be sprayed over Ma'amtagila Nation territory. Wilderness Committee. https://www.wildernesscommittee.org/news/toxic-glyphosate-be-sprayed-over-maamtagila-nation-territory.

Dawe, Charlotte. (2024). Toxic herbicide spray drops 97 per cent throughout Ma'amtagila territory in Great Bear Rainforest after alarm sounded. Wilderness Committee. https://www.wildernesscommittee.org/news/toxic-herbicide-spray-drops-97-cent-throughout-maamtagila-territory-great-bear-rainforest.

de Castro, C., and Lauer, A. (2025). The Gaia hypothesis revisited: Introducing an organic theory of Gaia, *EGUsphere,* preprint. https://doi.org/10.5194/egusphere-2025-1532.

Deur, D., Dick, A., Recalma-Clutesi, K., and Turner, N. J. (2015). Kwakwaka'wakw "clam gardens" motive and agency in traditional northwest coast mariculture. *Human Ecology* 43(2): 201–12.

Deur, D., Turner, N. J., Dick, C. C. A., et al. (2013). Kingcome Village's Estuarine Gardens as Contested Space. *BC Studies* 179: 13–37. Subsistence_and_Resistance-libre.pdf.

Dickson-Hoyle, S., Copes-Gerbitz, K., Hagerman, S. M., and Daniels, L. D. (2024). Community forests advance local wildfire governance and proactive management in British Columbia, Canada. *Canadian Journal of Forest Research* 54: 290–304.

Dickson-Hoyle, S., Ignace, R. E., Ignace, M. B., et al. (2022). Walking on two legs: A pathway of Indigenous restoration and reconciliation in fire-adapted landscapes. *Restoration Ecology* 30(4): e13566.

Donovan, Moira. (2022). Giving Up Glyphosate. *Maisonneuve.* maisonneuve.org/article/2022/01/25/giving-glyphosate/.

Filotas, E., Parrott, L., Burton, P. J., et al. (2013). Viewing forests through the lens of complex systems science. *Ecosphere* 5: art1.

Goodall, Jane. (1993). Chimpanzees: Bridging the Gap. In *The Great Ape Project,* edited by Paola Cavalieri and Peter Singer. St. Martin's Press. animal-rights-library.com/texts-m/goodall01.htm.

Grenz, J., and Armstrong, C. G. (2023). Pop-up restoration in colonial contexts: Applying an indigenous food systems lens to ecological restoration. *Frontiers in Sustainable Food Systems* 7: 1244790.

Groundwater, Jen. (2023). Stop the Spray: Glyphosate, Forest Monoculture, and Fire. *Watershed Sentinel.* https://watershedsentinel.ca/article/stop-the-spray/.

Heineman, J. L., Simard, S. W., Sachs, D. L., and Mather, W. J. (2005). Chemical, grazing, and manual cutting treatments in mixed herb-shrub communities have no effect on interior spruce survival or growth in southern interior British Columbia. *Forest Ecology and Management* 205: 359–74.

Heineman, J. L., Simard, S. W., Sachs, D. L., and Mather, W. J. (2008). Trembling aspen removal effects on lodgepole pine in southern interior British Columbia: Tenth year results. *Western Journal of Applied Forestry* 24: 17–23.

Henriksson, N., Marshall, J., Högberg, M. N., et al.(2023). Re-examining the evidence for the mother tree hypothesis—Resource sharing among trees via ectomycorrhizal networks. *New Phytologist* 239(1): 19–28.

Hosgood, Amanda Follett. (2024). Two BC Parties "Playing Catch-Up" with Glyphosate Pledges. *The Tyee.*

Ibarra, J. T., Cockle, K., Altamirano, T., et al. (2020). Nurturing resilient forest biodiversity: Nest webs as complex adaptive systems. *Ecology and Society* 25(2): 27.

Kopeky, A. (2024). "It's path-breaking": British Columbia's blueprint for decolonization. The Guardian. theguardian.com/world/2024/oct/09/british-columbia-blueprint-decolonisation.

Kuraś, P. K., Weiler, M., and Alila, Y. (2008). The spatiotemporal variability of runoff generation and groundwater dynamics in a snow-dominated catchment. *Journal of Hydrology* 352(1–2): 50–66.

Lazcano, A., and Pereto, J. (2017). On the origin of mitosing cells: A historical appraisal of Lynn Margulis endosymbiotic theory. *Journal of Theoretical Biology* 434: 80–87.

Li, X., Li, Y., Zhang, J., et al. (2020). The effects of forest thinning on understory diversity in China: A meta-analysis. *Land Degradation & Development* 31: 1225–40.

Lovelock, J. (2016). *Gaia: A New Look at Life on Earth*. Oxford University Press.
Ma'amtagila Nation. Little Big House. https://www.maamtagila.ca/little-big-house.
Ma'amtagila Nation. Our Territory. https://www.maamtagila.ca/our-territory.
Manilla, Desiree. (2022). "A Place to Make Things Right." *Watershed Sentinel*. https://watershedsentinel.ca/article/maamtagila-building-a-place-to-make-things-right/.
Mapes, Lynda V. (2022). On Vancouver Island, Land Back Looks Like Going Home. Atmos. https://atmos.earth/vancouver-island-land-back-indigenous-maamtagila/.
Mapes, Lynda V. (2025). *The Trees Are Speaking: Dispatches from the Salmon Forests*. University of Washington Press. https://www.jstor.org/stable/jj.27401526.
Margulis, L., and Chapman, M. J. (1998). Endosymbioses: Cyclical and permanent in evolution. *Trends in Microbiology* 6(9): 342–45.
'Namgis Forest Planning. (2017). Engage 'Namgis. https://engage.namgis.bc.ca/namgisforestplanning.
Perry, D. A., Amaranthus, M. P., Borchers, J. G., et al. (1989). Bootstrapping in ecosystems: Internal interactions largely determine productivity and stability in biological systems with strong positive feedback. *BioScience* 39: 230–37.
Pham, H. C., and Alila, Y. (2024). Science of forests and floods: The quantum leap forward needed, literally and metaphorically. *Science of the Total Environment* 912: 169646.
Pither, J., Pickles, B. J., Simard, S. W., et al. (2018). Below-ground biotic interactions moderated the postglacial range dynamics of trees. *New Phytologist* 220: 1148–60.
Rillig, M. C., Lehmann, A., Lanfranco, L., et al. (2024). Clarifying the definition of common mycorrhizal networks. *Functional Ecology* 396(6): 1411-1417.
Robinson, D. G., Ammer, C., Polle, A., et al. (2023). Mother trees, altruistic fungi, and the perils of plant personification. *Trends in Plant Science* 29: 20–31.
Ryan, T. (S.) L. (2014). *Territorial jurisdiction: The cultural and economic significance of eulachon* Thaleichthys pacificus *in the north-central coast region of British Columbia*. PhD dissertation, University of British Columbia. http://hdl.handle.net/2429/46515.
Sagan, L. (1967). On the origin of mitosing cells. *Journal of Theoretical Biology* 14(3): 225–74.
Simard, S. W., Hagerman, S. M., Sachs, D. L., et al. (2005). Conifer growth, *Armillaria ostoyae* root disease and plant diversity responses to broadleaf competition reduction in temperate mixed forests of southern interior British Columbia. *Canadian Journal of Forest Research* 35: 843–59.
Simard, S. W., Heineman, J. L., Mather, W. J., et al. (2001). Effects of operational brushing on conifers and plant communities in the southern interior of British Columbia: Results from PROBE 1991–2000. Land Management Handbook No. 48. British Columbia Ministry of Forests.
Simard, S. W., Heineman, J. L., and Youwe, P. (1998). Effects of chemical and manual brushing on conifer seedlings, plant communities and range forage in the southern interior of British Columbia: Nine year response. Land Management Report 45. British Columbia Ministry of Forests.
Simard, S. W., and Martin, K. (2012). The powerful networks of forests. *Friends of Clayoquot Sound Newsletter* Summer: 6.
Simard, S. W., Martin, K., Vyse, A., and Larson, B. (2014). Meta-networks of fungi, fauna and flora as agents of complex adaptive systems. In *Managing World Forests as Complex Adaptive Systems: Building Resilience to the Challenge of Global Change*, edited by K. Puettmann, C. Messier, and K. D. Coates. Routledge, 133–64.
Simard, S. W., Roach, W. J., Miller, E. M., and Beresford, H. (2025). Thinning to reduce

wildfire risk protects carbon stocks and increases understory plant diversity but simplifies stands in south coastal rainforests of British Columbia. In review.

Simard, S. W., Ryan, T. (S.) L., and Perry, D. A. (2025). Opinion: Response to questions about common mycorrhizal networks. *Frontiers in Forests and Global Change* 7: 1512518.

Simard, S. W., and Vyse, A. (2006). Trade-offs between competition and facilitation: A case study of vegetation management in the interior cedar-hemlock forests of southern British Columbia. *Canadian Journal of Forest Research* 36: 2486–96.

Steidl, James. Information on herbicide spraying in BC's Central Interior. Stop the Spray B.C. https://stopthespraybc.com.

Turner, N. J. 2008. *The Earth's Blanket: Traditional Teachings for Sustainable Living.* University of Washington Press.

White, E. (2006). Heiltsuk stone fish traps: Products of my ancestors' labour. Master's Thesis. Department of Archaeology, Simon Fraser University. https://summit.sfu.ca/item/4240.

Wickham, S. B., Augustine, S., Forney, A., et al. (2022). Incorporating place-based values into ecological restoration. *Ecology and Society* 27: 32.

Wood, L. J. (2019). The presence of glyphosate in forest plants with different life strategies one year after application. *Canadian Journal of Forest Research* 49(6): 586–94.

Zhang, H., Liu, S., Yu, J., et al. (2024). Thinning increases forest ecosystem carbon stocks. *Forest Ecology and Management* 555: 121702.

Zhou, L., Cai, L., He, Z., et al. (2016). Thinning increases understory diversity and biomass, and improves soil properties without decreasing growth of Chinese fir in southern China. *Environmental Science and Pollution Research* 23: 24135–50.

15: BACK TO THE GARDEN

Amazon Aid. Amazon and Water. amazonaid.org/resources/about-the-amazon/the-amazon-and-water/.

Authier, L., Violle, C., and Richard, F. (2022). Ectomycorrhizal networks in the Anthropocene: From natural ecosystems to urban planning. *Frontiers in Plant Science* 13: 900231.

Avital, S., Rog, I., Livne-Luzon, S., et al. (2022). Asymmetric belowground carbon transfer in a diverse tree community. *Molecular Ecology* 31, 3481–95.

Bachelot, B., Uriarte, M., McGuire, K. L., et al. (2017). Arbuscular mycorrhizal fungal diversity and natural enemies promote coexistence of tropical tree species. *Ecology* 98: 712–20.

Bennett, B. (1992). Plants and people of the Amazonian rainforest. *BioScience* 42(8): 599–607.

Bennett, J. A., Maherali, H., Reinhart, K. O., et al. (2017). Plant-soil feedbacks and mycorrhizal type influence temperate forest population dynamics. *Science* 355: 181–84.

Butler, R. A. (2012). Big trees, like the old-growth forests they inhabit, are declining globally. *Mongabay.* January 26, 2012. news.mongabay.com/2012/01/big-trees-like-the-old-growth-forests-they-inhabit-are-declining-globally/.

Cahanovitc, R., Livne-Luzon, S., Angel, R., and Klein, T. (2022). Ectomycorrhizal fungi mediate belowground carbon transfer between pines and oaks. *ISME Journal* 16: 5.

Cardoso, D., Särkinen, T., Alexander, S., et al. (2017). Amazon plant diversity revealed by a taxonomically verified species list. *Proceedings of the National Academy of Sciences* 114(40): 10695–700.

Delavaux, C. S., LaManna, J. A., Myers, J. A., et al. (2023). Mycorrhizal feedbacks influence global forest structure and diversity. *Communications Biology* 6(1): 1066.

Haq, H. U., Hauer, A., Singavarapu, B., et al. (2025). The interactive effect of tree mycorrhizal type, mycorrhizal type mixture and tree diversity shapes rooting zone soil fungal communities in temperate forest ecosystems. *Functional Ecology,* 39(6): 1441–54.

Horton, T. R., ed. (2015). *Mycorrhizal Networks,* vol. 224. Springer.

Klein, T., Rog, I., Livne-Luzon, S., et al. (2023). Belowground carbon transfer across mycorrhizal networks among trees: Facts, not fantasy. *Open Research Europe* 3: 168.

Klein, T., Siegwolf, R. T. W., and Körner, C. (2016). Belowground carbon trade among tall trees in a temperate forest. *Science* 352: 342–44.

Kuyper, T., and Jansa, J. (2023). Arbuscular mycorrhiza: Advances and retreats in our understanding of the ecological functioning of the mother of all root symbioses. *Plant and Soil* 489: 1–48.

Mazzocchi, F. (2012). Complexity and the reductionism-holism debate in systems biology. *WIREs Syst. Biol. Med.* 4: 413–27.

McGuire, K. L., Henkel, T. W., Cerda, I. G. D., et al. (2008). Dual mycorrhizal colonization of forest-dominating tropical trees and the mycorrhizal status of non-dominant tree and liana species. *Mycorrhiza* 18, 217–22.

McMaster, G. (2020). *Iljuwas Bill Reid: Life and Work.* Art Canada Institute. aci-iac.ca/art-books/iljuwas-bill-reid/significance-and-critical-issues/.

Merckx, V. S. F. T., Gomes, S. I. F., Wang, D., et al. (2024). Mycoheterotrophy in the wood-wide web. *Nature Plants* 10: 710–18.

Mills, M. B., Malhi, Y., Ewers, R. M., et al. (2023). Tropical forests post-logging are a persistent net carbon source to the atmosphere. *Proceedings of the National Academy of Sciences* 120(3): e2214462120.

Morin, B. (2025). Indigenous guardians successfully keep extractives out of Ecuador's Amazon forests. https://news.mongabay.com/2025/10/indigenous-guardians-successfully-keep-extractives-out-of-Ecuador's-Amazon-forests/.

Pachamama Alliance. The Achuar. https://pachamama.org/achuar.

Pachamama Alliance. (2021). Bio-entrepreneurship in the Amazon. news.pachamama.org/bio-entrepreneurship-in-the-amazon.

Peh, K., Lewis, S. L., and Lloyd, J. (2011). Mechanisms of monodominance in diverse tropical tree-dominated systems. *Journal of Ecology* 99, 891–98.

Robbins, J. (2021). How Returning Lands to Native Tribes Is Helping Protect Nature. *Yale Environment 360.* e360.yale.edu/features/how-returning-lands-to-native-tribes-is-helping-protect-nature.

Robinson, J. M., Gellie, N., MacCarthy, D., et al. (2021). Traditional ecological knowledge in restoration ecology: A call to listen deeply, to engage with, and respect Indigenous voices. *Restoration Ecology* 29: e13381.

Rowell, A. (2017). *Green Backlash: Global Subversion of the Environmental Movement.* Routledge.

Ter Steege, H., Pitman, N. C., Sabatier, D., et al. (2013). Hyperdominance in the Amazonian tree flora. *Science* 342(6156): 1243092.

Veit, P., Gibbs, D., and Reytar, K. (2023). Indigenous Forests Are Some of the Amazon's Last Carbon Sinks. World Resources Institute, January 6, 2023. wri.org/insights/amazon-carbon-sink-indigenous-forests.

Washington Post. (2022). Amazon deforestation hits new record in Brazil. washingtonpost.com/climate-environment/2022/07/08/amazon-rainforest-deforestation-record-climate/.

Wright, J. S., Fu, R., Worden, J. R., et al. (2017). Rainforest-initiated wet season onset over the southern Amazon. *Proceedings of the National Academy of Sciences* 114 (32): 8481–86.

YVR. The Heart of the Airport: *The Spirit of Haida Gwaii: The Jade Canoe.* https://www.yvr.ca/en/about-yvr/art/the-heart-of-the-airport.

Zahra, S., Novotny, V., and Fayle, T. M. (2021). Do reverse Jansen-Connell effects reduce diversity? *Trends in Ecology & Evolution* 36: 387–90.

CONCLUSION: FINDING OUR WAY FORWARD

Blicharska, M., and Mikusinski, G. (2014). Incorporating social and cultural significance of large old trees in conservation policy. *Conservation Biology* 28: 1558–67.

Drever, C. R., Cook-Patton, S. C., Akhter, F., et al. (2021). Natural climate solutions for Canada. *Science Advances* 7(23): eabd6034.

Franklin, J. F., Johnson, K. N., and Johnson, D. L. (2018). *Ecological Forest Management.* Waveland Press.

Moomaw, W. R., Masino, S. A., and Faison, E. K. (2019). Intact forests in the US: Proforestation mitigates climate change and serves the greatest good. *Frontiers in Forests and Global Change* 2: 27.

Ripple, W. J., Wolf, C., van Vuuren, D. P., et al. (2024). An environmental and socially just climate mitigation pathway for a planet in peril. *Environmental Research Letters* 19:021001.

INDEX

Page numbers in *italics* refer to illustrations.
Abbreviation: MTP = Mother Tree Project

abundance, 9, 18
Achuar people, 218–28, *225*
Adam and Eve Rivers, 204
aggregated retention system, 36
Alaska, 237
alders, 16, 79, 189
    red, 187
    Sitka, 165
Alert Bay, 184, 204
Alert Bay residential school, 183
Alex Fraser Research Forest, *34*, *35*, 59–60
allowable cut rate, 37–38
alluvial fans, 124
aluminum, 91
Alzheimer's, 71–75, 95
Amazon rainforest, 11, 218–27, *225*, 231
Amazon River, 218
Ambers, Matthew, 185, *187*
Ancient Forest Alliance, 180
Andes Mountains, 228
ant, leafcutter, 221, 226
anticarcinogens, 102
arbuscular mycorrhizal fungi, 101, 104, 117–18, 157, 167, 222
Arctic Circle, 231
Armillaria root disease, 21
Armstrong, Gaelin, 148
arnicas, heart-leaved, 16, 152
A.R. Rogers Sawmill, *87*
Arrow Lake, 59, 198

arthropods, 6, 56
arútam (forest spirit), *225*, 226–27
Asay, Amanda, 28–32, *31*, 97–100, 104–6, 157–59, 164, 168, 186, 194–97, *196*, 209, 216–17
    death of, 194–97, 201, 213–14, 221, 229
Asay, George, 196
aspen, trembling, 60, 67, 69–70, 78, 99, 146–48, 154, 164–65
Athabascan people, 106
atmosphere, 18, 20, 165, 218
ATP, 58
autumn litter fall, 164–65
Awi'nakola Foundation, 203

bacteria, 3, 57, 91
badger, 66
balsa, 220
barnacles, 89, 185
Barriere, British Columbia, 72
Baseball America, 29–30
bats, 165, *225*
BBC, 213
Beard (forest defender), 171
bears, 6, 37, 50–51, 78, 85, 120, 122, 165, 188–89, 204
    clearcuts and, 130
    dens inside ancient trees, 90, 175, 177, 183, *184*, 228

bears *(continued)*
    Mum and, 15
    salmon and, 89–93
bear species
    black, 53, 111, 175, 182, *184*
    grizzly, 20, 53, 60, 63, 89–91, 141, 155, 182, 185, 228
Beauregard, Jacob, 145–46, 153
beetle, 165, 221
    Douglas-fir bark, 38, 70
    mountain pine, 38, 70
    spruce-bark, 38
beetle infestations, 70, 166, 210, 231
Before-After-Control-Impact (BACI), *62*
Bella Bella, British Columbia, 82, 91–92
Binche Whut'en territory, 60
biodiversity, 9, 107, 166, 229
    carbon and, 18
    clearcuts and, 7, 122, 131, 205
    Haida Gwaii and, 121–22
    Indigenous stewardship and, 188
    Kwiakah and, 182
    Ma'amtagila and, 205
    mixing of trees and, 165
    mother trees and, 171
    MTP measures of, *62*
bioelectrical waves, 235
biofuels, 136
biomass, 18
birch, paper, 69, 73, 81, 97–98, 147–48, 165, 168
birds, 8, 165, 188–89, *225*
blueberry, 7, 205
    lowbush, 198
    oval-leaved, 114
bogs, 113–15
bolete, king, 205
Bonaparte River watershed, 64–66
Box Lake, 161
Box Mountain, 160–61
breast cancer, 4, 93, 102, 161, 202
Britain, 22, 100
British Columbia Green Party, 237
British Columbia Ministry of Forests, 44, 108, 112, 158–61, *162*
British Columbia provincial government, 37, 38, 42, 68, 70, 108–9, 112, 115, 121–24, 170–81, *179*, 236

British Columbia Sports Hall of Fame, 201
British Columbia Supreme Court, 236
broadleaf trees, 166
bromeliads, 220
Brown University, 29
brush fields, 154
brush saws, 69
bud banks, 206
budworm, western spruce, 27, 38
buffaloberry, 75
burn, low-intensity cultural, 7, 36
Burnaby Island, 108
burn piles, 119, 132
butterfly, swallowtail, 17

Cache Creek, British Columbia, 44, 64–66
Cake (forest defender), 170
calcium, 91, 164
California, 210
Calvin cycle, 58
cambium, 76, 103
Campbell River, 141, 204
Canada
    as British colony, 22
    greenhouse gas emissions and, 231–32
    Haida agreement and, 109
    logging quotas and, 85–86, 115
    Ts'msyen law and, 82
Canadian Interuniversity Sport hockey, 29
Canadian women's baseball team, 29–30, 97, 158, 201
canopy
    gaps in, 22, *31*, *35*, 36, 100–1, 161, 187, 226
    water cycle and, 77–78
capillary action, 226
capybara, 220
carbon
    Amanda's research on, 158
    Amazon and, 223–24, 226
    atomic structure and, 17–18
    birch and fir transmission of, 97–98
    bole-only vs. whole tree removal and, 136
    clearcutting and, 7, 122, 138, 141–43
    cryptogamic crust and, 149
    cycle of, 130, 218

dead logs and, 5, 18, *19*, 22
Eva's Septuplet study and, 105, 157–59, 167
forest diversity and, 165–66
global, 20, 191
Haida forests and, *111*, 113–15, 119, 121, 139–41
humus depths and, 15
Indigenous managed territories and, 222–23
isotope labeling to trace movement of, 97, 105, 168, 208–9
kin Douglas firs and, 30
logging and, 125–38
mosses and lichens and, 149
Mother Trees and, 172
MTP sites and, *62*, 125–28, 137–38, 153, 232–33
mycorrhizal fungal networks and, 4, 6
old-growth cedar-hemlock forest and, *129*
photosynthesis and, 17–18, 20
pricing of, *49*
restoring, to forests, 191
salmon-rich forests and, 91–93, 139
sequestration and storage of, 7, 16, 18, 78, 93, 165, 180, 229, 232
soil food web and, 5–6
understory vs. overstory, 23
water and, 78
Carbon Budget Model of the Canadian Forest Sector, 137
"Carbon Storage and Biodiversity Across a Climate Gradient in Interior Douglas-Fir Forest" (Simard and Roach), 125
Cariboo Mountains, 32, 64, 162
Cariboo MTP site, 13–19, 21–25, 64, 75–79
caribou, southern mountain, 37, 148
Castlegar, British Columbia, 160, *162*
cattle, 152
cavitation, 224
cavity nesters, 6
Cayuse Creek watershed, 174, 180
CBC, 73, 213
cedar, 49, 94, 100, 112–21, 146, 151, 161–65, 205
arbuscular mycorrhizal fungi and, 101
bear dens inside, 90, 175, 183, *184*

Eva's Septuplet study of yew, Douglas fir, and maple and, 103–5, 117, 157–59, 167–68, 172
Fairy Creek and, 170
germinants, 148
grandmother, 115–16, 186
Haida and, 112–24, *121*, 191
harvesting, to weave into baskets, 79–81, *81*, 103, 109
hemlock and, *140*, 141–42, *190*
kin relationships and, 186
Kwakwaka'wakw and, 183–86
layering and, 116–17
Malcolm Knapp Research Forest and, 83, 112
MTP clearcut treatments and, 149
Nimmo Bay Mother Tree, 89–90
nitrogen and, 165
salmon and, 92
Selkirk Mountains and, 123–24
totem pole (*gyaa'ang*), 120
cedar species
western redcedar, 54, 80, 85, *111*, 112–19, *121*, *129*, 147, *184*, 191
yellow, 118–19, 169, 174–77, *178*
chanterelle, golden, 205
Charles II, King of England, 37
chemotherapy, 93, 102, 211, 228
cherry
bitter, 161
choke cherry, 187
wild, 81
chestnut, 68
chickadee, black-capped, 76
chimpanzees, 216
China, 88
Chiolerio, Alessandro, 233
chlorophyll, 18
Christian missionaries, 22, 120
clam, 84, 86–87, 207, 235
clam garden, 87–89
clam shell middens, 87, 187
Clayoquot Sound, 176
clearcutting, 6, 10, 135, 166
bears and, 60, 91
British Columbia provincial law on, 170
Cache Creek flooding and, 65–66
carbon losses due to, 126–38, 140, 142, 176, 234

clearcutting *(continued)*
  caribou and, 148
  clearcuts with reserves vs., 128–29, *131*
  defined, *34*
  Fairy Creek and, 168–77, *178*
  greenhouse gases and, 119, 233
  Haida cedars and, 117–19, *118*
  Haida Gwaii and, 107–11, *110*, *111*, 114–15, 121–24, *123*, 140
  history of, in British Columbia, 37–40
  Indigenous opposition to, 44, 46
  invasive plants and, 149
  Kwagu'l territory and, *187*
  Kwiakah territory and, *189, 192*
  Ma'amtagila territory and, 183, 184, 203–7
  MTP sites and, 7, 32, 111, 121–22, 126, 148–50
  need to halt, 138
  Nimmo Bay, 89, 90
  prevalence of, in British Columbia, *34*, 128–30, 150
  profits and, 128
  protests vs., 160
  reforestation and, 154
  regenerating seedlings and, 233
  restoration and regeneration after, 205–6
  Selkirk Mountains and, 123–24
  slash piles and, *49*
  snowmelt and, 124
  soil organo-mineral compounds and, 137
  species diversity and, 233
  wildfires and, 68
climate change, 7, 9, 27, 32–33, 37–39, 67, 70, *111*, 122–23, *123*, 133–39, 155, 229, 232
Clostridioides difficile, 160, 197
cloudberry, 114, 205
clover, springbank, 87, 207
coastal forest, 64, 80, 88, *111*, 121–22, 135. *See also* specific locations
  MTP sites, 83, 139–40
cockroach, 221
co-dominants, 22
coevolution, 4
Cog (forest defender), 171
collembola, 57, 177
colonization, 9, 22
Columbia Mountains MTP site, *129*
Columbia River, 96, 219
community-based management, 110
conifers
  needles of, 77
  plantations of, 189, 205
  subsurface water and, 166
  wildfires and, 69–70
Conroy, Katrine, 160
Constantinou, Alexia, *62*, 154
continuous cover forestry, 36
Cookie (forest defender), 174, 176
Cook-Makwala, Rande, 173, 182–87, *184*, 191–93, 203, 205–7, 209–10, 216
Cook's Ferry Indian Band, 42
COP26 conference (Glasgow), 131
Cora (Indigenous girl), 112–13
Cornelis, J.T., 137
cottonwood, black, 147, 164–65, 198–99, 205
Cottonwood Bay, 199
cougar, 154
covalent, 17
COVID-19 pandemic, 144, 150, 194
Cowichan estuary, 207–8
coyote, 67, 154
crabapple, 88, 187
cranberry
  bog, 114
  highbush, 187
Cranbrook, British Columbia, 59
crane, sandhill, 114
crowberry, 114
cryptogams, 149, 154

Daajing Giids, British Columbia, 110
Dakelh territory, *19, 145*
Da'naxda'xw Awaetlala Nation, 171
dandelion, common, 149
dead wood pool, *19*
deciduous trees, 69–70
decomposition, 5
deer, 148, 188, 204
  mule, 67
  Sitka black-tailed, 115, 117, *118*
  white-tailed, 152, 154
deforestation, 131
devil's club, 189, 205, 207

INDEX

Dhanwant, Kiku, 206
Dick, Mike, 141
Dick, Steven, 141
Ditidaht Nation, 180, 237
DNA, 58, 117
dolphin, pink, 218, 220
dominant trees, 9, 22
Douglas fir, 60, 65, 185, 208
   absence of, on Haida Gwaii, 112
   Amanda's studies on, 158, 202
   canopy gap and, *31*
   carbon transmission between birch and, 98, 168
   clearcutting and, 98–99
   connections between older and younger, 164, 186
   drought and, 27
   ectomycorrhizal fungi and, 101
   Eva's Septuplet study and, 103–5, 159
   fire and, 67–68, 75–76
   genotypes of, moving northward, 54–56
   glaucous sheath of needles and, 18
   Jaffray site and, 150–51
   kinship and, 29–30, *31*, 36, 98–99, 186, 202, 209, 222
   larches and pines and, 101
   Malcolm Knapp Research Forest site and, 83
   morels and, 78
   mother trees and, 17, 21–22, 99, 113
   MTP sites and, 8, 11
   plantations and, 188–89
   regeneration of, *31*, 147
   Selkirk Mountains and, 123–24
Douglas fir species
   coastal, 55, 83
   interior, 8, *31*, 54–55, 83, 98, 112, 125, *145*, 151
dragonfly, 216
drought, 9, 99, 188, 210, 224, 234
   of early 2000s, 38–39
   of 2021, 159
   of 2022, 214
   wildfires of 2017 and, 67–68

eagle, 20, 85–86
   bald, 90, 185
ecotourism, 223
ectomycorrhizal fungi, 50, 101, 152, 164, 222
Ecuador, 218, *225*
Edgewood, British Columbia, 21, 198
elderberry, 91, 187, 205
   blue, 189
   red, 88
electrome, 234
electron transport chain, 58
Elephant Hill fire, 64–68, *66*, 231
Elephant Mountain, 208, 210
elk, Rocky Mountain, 152, 154
endangered species, 174–75
Enderby, British Columbia, 200
endophytes, 222
endosymbiosis, 216
Enlightenment, 210
enzymes, 58
ephiphytes, 220
ericoid mycorrhizas, 117
ermine, 110, 204
eskers, 113–15
ESSA, 180
eucalyptus, 68–69
eukaryotic cells, 216
European Forest, 15
evapotranspiration, 76–77
Extinction Rebellion, 160

fairybell
   Hookers, 169
   largeflower, 174
Fairy Creek, 169–80, *178*, *179*, 228
   agreement of 2021, 180–81
   agreement of 2024, 236–37
Fairy Creek forest defenders, 10, 160, *162*, 169–80, *179*, 229, 236
false azalea, 149
far-red light, 164
feller buncher, 40, 127, *136*, 233
fern, 142, 172
   lady, 79, 118
   licorice, 164
   northern maidenhair, 164
   oak, 149, 163
   spiny wood, 164
fig, strangler, 220
Finding the Mother Tree (Simard), 4, 208

298 INDEX

fine fuels, 21
fire, benefit to ecosystems, 67. *See also* burn; wildfires
fireweed, 57, 59, 67, 75, 146, 214
Fireweed (forest defender), 171
fir, 97, 162. *See also* Douglas fir
    grand, 147–48
    Pacific silver, 174
    subalpine, 214–15
First Nations (Indigenous peoples) 7, 182–93. *See also* Indigenous stewardship; *and specific groups*
    agrarian and marine technologies of, 88–89
    Amazon rainforest and, 219–26
    ancestral knowledge of, 93–94
    carbon-rich territories of, 222–23
    cedars and, 89
    connecting with, 9, 27
    cosmology of, 82
    Dave Walkem as first registered forester, 41–42
    dispossession of, 22, 37, 112–13, 182
    Fairy Creek and, 172–73, 178–81
    fire stewardship and, 22–23, 70, 85
    forest restoration and, 182–83, *189*, 191–93, *192*
    forest stewardship, 7, 83
    Hudson's Bay Company and, 22, 37
    kinship and, 98
    Land Back movement and, 235
    land management and, 222–23, 235–36
    learning from, as guardians of natural world, 235
    logging on tenures of, 43, 172, 180–81
    medicine wheel and, 17
    mother trees retained by, 155–56
    MTP and, 8, 45–46, 83, 119
    residential schools and, 22, 26–27, 112–13, 183
    restrictions on, 85
    return of ancestral lands to, 235–36
    as small-time loggers, 42
    stewardship of resources and, 81, 182, 188
    tenures withheld from, 42–43
    traditional crafts and, 79–81, *81*
    values of, 5, 46, 134, 156
    yew's medicinal properties and, 102

fish, yews and, 100. *See also* salmon
fisher, 204
fisheries science, 82
fish farms, 204, 235
fishing, ancient technologies, 83–85, 207, 235
flickers, 6
floods, 37, 39, 63–66, 124, 231
    forest diversity and, 165
    Indigenous stewardship and, 188
    reducing risk of, 70
foamflower, threeleaf, 79
food chain, 6, 20, 207
    keystone species and, *111*
forests. *See also* clearcutting; logging; Mother Tree Project; mother trees; old-growth forests; timber industry; *and specific concepts; locations; and species*
    bears and, 89–94, 130, 175, 177, *184*, 227
    carbon and, 6, 17–18, 77–78, 91, 93, 104, 115, 122, 182, 191
    carbon emissions from Canadian, 232
    carbon in floor of, and logging, 126–43, 232–34
    climate change and, 122–24, 133, 155–56
    cycle of birth, death, and renewal and, 3–6, 10–12, 229–30
    deciduous trees and vitality of, 69–70
    decline and degradation of, 27, 93, 115, 118–23, 160–61, 182–83, 191, 223, 229
    developing plans to restore clearcut, 183–93, *189*, *190*, 192, 203–7
    diverse, replaced by monoculture plantations, 68–70, 93, 142, 188–91
    diversity and interconnectedness of, 3–5, 20, 99, 105, 107–8, 116, 210–11, 237–38
    diversity of old-growth, 187–88
    future of, and learning to manage and restore, 231–38
    history of logging in British Columbia and, 37–39, 42–44
    holistic vs. reductionist view of, 210–11
    hydrology of ecosystem and, 113–14
    impact of wildfires and, 64–71, 75, 78
    Indigenous ceremony mourning loss of, *187*

INDEX 299

Indigenous people and, 5, 7, 42–47,
    79–85, 98, 102, 106–121, 141, 155–56,
    171–73, 178–93, 229, 234–37
Indigenous people of Amazon and,
    218–28
  kinship relations and, 29–30, 98–100,
    186–87, 213
  medicinal species and, 102–4
  mycorrhizal fungal networks and, 4,
    23–24, 78, 101, 105, 116, 119, 163–64,
    167–68, 172, 176–77, 186, 208–11,
    222
  niche complementarity and, 164–67
  oldest, or mother trees, as hub of, 4–5,
    23–24, 76, 98, 99, 105–6, 115–16,
    155–56, 171–72, 176–77, *178*, 183,
    185–86, 205–6, 209, 224–28, *225*,
    234–35
  overstory and protection of understory,
    105–6, 155, 167–68
  photosynthate transmission among
    species and, 100–5, 167–68
  photosynthesis and respiration and,
    18–20, 77–78, 163
  place matters concept and, 154–55, 167
  protest movements to defend old-
    growth, 108–9, 160–61, *162*, 169–81,
    179, 236–37
  provincial regulation of logging and,
    118, 121–24, 178–81, *178*, 236–37
  resilience and complexity of, 214–15,
    230, 237–38
  salmon and, 89–94, 122
  search for better ways to tend, and
    MTP design, 6–11, 39
  seedling migration and, 55–56, 154
  settlers and small loggers and, 85–89
  shifting from exploitation to
    stewardship of, 9, 210, 226–27
  SISCO conference on, 125–26, 131–32
  timber company profits and, 115
  transpiration and, 77
  value of standing, vs. logging, 180
  water cycle and, 93
  Western science and, 182, 193, 209–11,
    213, 216
  whole tree logging and, 13
forget-me-not, 211
forsythia, 73
Fort St. James, 56, 60
forwarders, 127, 233
fossil fuels, 223, 228, 231
fox, red, 90
Fraser, Kaya, 60, 62, 147
Frederick Arm, *189*
frogs, 225
fuel load, *19*, 27, 67
fungi, 3, 11, 58, 91, 162, *215*. See also
    mushrooms; mycorrhizal fungal
    networks; *and specific types*
  pathogenic, 102, 222

Gadget (forest defender), 171
Gagliano, Monica, 234
Gaia hypothesis, 216
gametophores, 75
gap dynamics, 195
gap function, 166
gastropods, 58
Georgia Strait, 157
gilia, scarlet, 67
ginger, wild, 187
gingko, 74
Giveout Creek, *136*
glacial moraines, 113, 115
glucose, 58
glycolysis, 58
glyphosate (Roundup), 189, 205
goat, mountain, 213
goldenrod, Canada, 154
Gold Rush, 44
Goodall, Jane, 216
goshawk, northern, 107, 110, *111*
Graham Island, 109–10, 120, *121*
grape, 74
  tall Oregon, 207
grapple skidder, *136*
grass, 149
  bear, 81
  canary, 81
  couch, 149
  English ryegrass, 149
  orchard, 152
  pinegrass, 45, 59, 67, 151
Great Bear Rainforest, 89, 176, 183–84,
    189
Greece, 210

300   INDEX

greenhouse gases, 38, 40, 70, 119, 130, 138, 231–33
greenwashing, 128
Grenz, Jennifer, 207
ground truthing, 191
Guichon, Shannon, 172
Gulf of Alaska, 83
Gwaii Haanas National Park Preserve and Haida Heritage Site, 109
Gwawaenuk Tribe, 89

habitat, 8, 18, 165
   destruction of, 37, *111*, 205
   indigenous stewardship and, 188
Hada, British Columbia, 184–85, 187, 191
Haida Gwaii, 107–22, *110*, *111*, *118*, 141, 191
   carbon and salmon-rich forest of, 139–40
   clearcuts and, 115–24, *118*, *121*, *123*, 130, 131, 140
   Haida Nation title ratified, 236
   hemlock saplings in second-growth forest, 206
   MTP and, 111–12, 121–22, 140
   non-native deer and, 115
   Teal-Jones Group lawsuit and, 181, 236
   timber supply review and, 122
Haida Gwaii Management Council (HGMC), 123, 181, 236
Haida Nation, 83, 102, 107–17, *110*, *111*, 120, 122, 235
   mythology of, 229
   rebellion of 1985 and, 108–9
   title to Haida Gwaii, 236
   totem poles and, 120–21
Hank (logging foreman), 40, 48, 51–52, 75, 76
harvesting licenses, 56
harvesting on short rotations, 166
haustoria, 163
hawks, 125
hawkweed, white, 51
hazelnut, beaked, 74, 161, 187
Headwall ski runs, 161
Heard Museum, 79
heat, extreme, 9, 39–40, 70
heat dome, 67, 159, 214, 231

Heath, Bill, 92–93, 96, 160
Heath, Kelly Rose, 20, 79, 96, 194, 197, *197*, 212
Heath, Oliver, 137
Hecate Strait, 107, 113
Heiltsuk Integrated Resource Management Department, 83
Heiltsuk Nation, *82*, 83–86, 91, 139
heliconia, 220
hemlock, 79, 89, 112, 117, 123, 151, 161–65, 185
   clearcuts and, 114
   ectomycorrhizal fungi and, 101
   germinants, 146, 148
   mycorrhizal fungal networks and, 163–64
   plantations and, 206
hemlock species
   mountain, 169
   western, 50, *111*, *129*, 149
herbicides, 69, 189, 205
herring, Pacific, 177
heterogeneity, 166
heterotrophic carbon, 163
heterotrophic food chain, 207
Hiellen River, 120
Hiladi village site, 204–7
Hišuk ma c̓awak Declaration, 180
hoe chucker, 135, 233
holistic stewardship, 193
Holmes, Jody, 141–42, 188
Holt, Magali, 194–95
Holt, Rachel, 141, 183–84, 194–95
honeysuckle, 13
Hong Kong, *14*
horse-logging, 85, *87*, 133
horsetail, common, 154
huckleberry, 21, 117, 163, 188–89, 205, 207
   Alaska, *19*
   black, *19*, 88
   blue, *19*
Huckleberry (forest defender), 171
Hudson's Bay Company, 22, 38
humus, 15–16, 113, 132, 139, 141, 176
Huu-ay-aht Nation, 180
Húy̓at village site, 87
hydrogen, 18
hydrology, 77–78, 113, 115, 172, 207

Indian Act (1876), 85, 172
Indian ice cream (*sxusem*), 21
Indigenous stewardship, 5, 7–9, 44, 70, 89, 92, *111*, 180, 187–88, *191*, 192, 232, 235–36. See also First Nations; *and specific groups*
infestations, 165, 205
insects, 20, 165, 220
    pests, 39
    pollinating, 188
interdependent web of nature, 93
interspecies collaboration, 166
intertidal food system, 208
invasive species, 149, 152, 188
isotopic labeling, 105, 157–58, 167–68, 187, 208–9
Italy, 234–35
Itsekin village site, 187–88

Jaffray MTP site, 59, 92, 125, 137, 146, 150–53
jaguar, 218
John Prince Research Forest MTP site, *19*, 55, 60–63, 125, 144–45, *145*, 153–54
Johnstone Strait, 203
Jones, Bill, 171–73, 180
Jones, Liam, 57
Joseph, Robert, 119
juniper, common, 152, 158

Kamloops, British Columbia, 48, 112
Kamloops Indian Residential School, 112–13
kapok, 223–26, *225*, 228, 230
Kelowna, British Columbia, 125, 131
Kilmer, Joyce, 211
Kimmins, Hamish, 135
Kinbasket Lake, 219
kinnikinnik, 207
kin relationships, 30, *31*, 98–100, 158, 186, 209, 213
Klamath River, 236
Knox, David, 173
Kootenay Lake, 15, 96
    fire of 2015 and, 73
    MTP site, 32, 53–59, 137, 144–51, 159

Krebs cycle, 58
Ktunaxa Nation, *16*, *49*, *114*, *129*, *136*, *162*
Kunghit Island, 109
Kunsoot estuary, 84, 93
Kunsoot River, *82*
Kuskanax River, 70, 198
Kwakiutl Nation, *140*, 141–43, 173
Kwakwaka'wakw peoples, 83, 102, 119, 139, 173, 183–86, 187, 203–4
Kwak'wala language, 185
Kwiakah Nation, 141, 182, *189*, *192*, 235
Kwikwasut'inuxw Haxwa'mis Nation, 183

Lakeside Park, *16*, 73
Lana (Indigenous girl), 112–13
Land Back movement, 235
land degradation, 130–31, 133–34
landslides, 66, 71, 124
Lantau Island, *14*
larch, western, 54, 56–57, 101, 148, 151, 162
Laroque, Allen, 91–93
laser hypsometer, 86
laurel
    Portugal, 68
    sticky, 75
Lavkulich, Les, 8, 137
Law, Danica, 144, 146, 148, 152
layering process, 116–17, 163, 185
Le Guin, Ursula K., 189
lettuce, prickly, 51
lianas, 220
lichen, *19*, 75, 141, 161, 163, 165, 171, 189
    coral, 149
    dog, 50
    fairy puke, 169
    horsehair, 148
    lung, 165
    old-growth specklebelly, 174
    pelt, 152, 173
    reindeer cup, 152
    witch's hair, 148
Lichen (forest defender), 171
lightning, 67–68, 159
Likely, British Columbia, 13
lilac, 74
Lillooet Mountains, 14
lily, 163
    avalanche, 50, 52

liverwort, 19, 67, 141, 163, 171
  lime-green, 75
log drives, excessive, 87
logging, 229. *See also* clearcutting; timber industry
  carbon and, 18, 49, 125–38, 140
  Fairy Creek injunctions and, 178–79
  forest regeneration after, 11–12
  greenhouse gases, 130
  Haida rainforest and, 107
  MTP patterns and, 8, 41–42
  opposition to industrial, 204
  percent of, as product, 49
  quotas and, 115
  rivers and, 122–23
  stopping, in primary and old-growth forest, 234
logging equipment, impact of, on forest floor, 126–27, 133–37, 136, 140, 152, 169, 233–34
logging roads, 70, 118, 172, 189–90
logs, decaying, 3, 5, 16, 22, 228
Lovelock, James, 216
lungwort, 174
lupine, large-leaved, 164–65
LVR Secondary School, 72
Lyell Island, 108–9
  blockade of 1985, 229
lymphoma, 102
Lytton, British Columbia, 44

Ma'amtagila Nation, 173, 183–90, 184, 203–8, 235
  recognition of rights and title and, 183
  restoration of forest and, 205–8
Mabel Lake, 37, 43, 85–86, 87, 103, 198–99, 220
macaw, scarlet, 218, 220–21
Macbeth (Shakespeare), 101
M'ac'inuxw Special Management Area, 189
magnesium, 18, 164
mahogany, 220
MAID (medical assistance in dying), 95–97, 159–60, 197–98
Malcolm Knapp Research Forest MTP site, 83, 92, 112, 125, 137, 140, 141, 191

manganese, 91
Maori pounamu, 113
Mapes, Lynda, 173–75
maple, 125, 164, 189, 228
  arbuscular mycorrhizal fungi and, 101
  bark, weaving, 81
  Eva's Septuplet study on yew, Douglas fir, cedar and, 104–5, 117, 159, 167
  MTP clearcut treatments and, 149
maple species
  bigleaf, 80, 187
  Douglas, 15, 54, 100–1, 161
Margulis, Lynn, 216
marmot, 214
marten, 90
Martin, Ron, Jr., 84, 86
Massett Inlet, 110, 114
medicinal plants, 207, 223
medicine wheel, 17
Mediterranean woodlands, 11
Merkel, Garry, 180–81
metabolic pathways, 164
metagenomic techniques, 91
Mexico, 56, 83
microbes, 77
microbial biofilms, 177
microbiomes, 164
Mills family, 112
mineral industry, 223
mineral nutrients, morels and, 78
mineral soil layer, 15, 127, 136, 140, 142, 149, 152, 164
Mission Ranch, 195
mites, 58
Moe (forest defender), 170–71
Mohn, Bill, 8
monkey
  brown-headed spider, 221
  capuchin, 221
  mantled howler, 221
monoculture plantations, 68–69, 93, 142, 166, 188–92, 190, 205
monoterpenes, 77
Moore, Keith, 236
moose, 67, 188
Moreira, Vital, 68–69
morel, 67
Moresby Island, 108, 110
mosquitoes, 19

moss, *19*, 75, 107, *111*, 113, 140, 159, 171, 189, 204
  bristly haircap, 152
  broom forkmoss, 152
  electrified cats-tail, 50
  large leafy, 142, 149
  Oregon, 149
  pipecleaner, 169
  purple horn-tooth, 152
  red-mouthed leafy, 50, 149
  red-stemmed feathermoss, 116, 152, 163
  slender beaked, 142
  sphagnum, 114
  step, 21, 116, 149
Mother Tree Project (MTP), 219, 228. *See also specific sites*
  assessments of, in 2020, 144–55
  Awi'nakola Foundation and, 203
  carbon stocks and, 92, 142, 166, 232–34
  Cariboo Mountains plot, 13
  clearcut plots, 33, *34*, 40, 47, 147–51
  climate gradient and, 33, 83
  concept, design, and goals for, 7–11, 40, 83, 237
  control plots, untouched, 36–37
  crew and, 10, 32, *62*, 190–91
  depth of forest floor at Phillips Arm vs. interior forests, 141–42
  Douglas fir genotypes and, 55
  drought and, 224
  early results in clearcuts vs. partial cuts, 125–31
  Eva's Septuplet study and, 104, 159
  experimental sites of 2018, 122
  forest floor carbon findings and SISCO, 125–39, *129*
  funding and, 8, 112
  Hada site proposed, 191
  Haida Gwaii and, 111–12, 119–22, 140
  harvesting and treatment design and, 32–33, 36–38, 41–42, 48, 51–53
  Heiltsuk plot and, 83–85
  Indigenous knowledge and, 8–9, 43, 45–47
  Jean Roach as manager and, 8
  Kwiakah as partner and, 141–42
  Mother Trees and, 33–36, *34*, 41–42, 234–35
  naturally regenerated seedlings and, 147
  planting and, 10, 49, 53–63, *62*
  regenerative forest stewardship and, 83
  retention design and methods, 33, 46–47
  retention plots, ten percent or seed tree, 33–36, *34*, 41–42, 51–53, 128, 147–52
  retention plots, thirty percent, *35*, 36, 41–42, 128, 147, 151–52, 154
  retention plots, sixty percent, *35*, 36, 39–40, 50, 59–60, 127, 146–47, 150–53
  seedling genotypes and, 54–55
  Twaal Creek site and, 47
  website for, 238
mother trees
  carbon and, 23
  cedar layering and, 116–17
  clearcutting and, 99
  concept of, as metaphor, 42, 93, 209
  decaying logs and, 5–6
  disappearance of, 189
  diversity of connections of, 227–28
  Douglas fir and, 115
  Haida Gwaii cedar and, *111*, 112, 115–16
  importance of, for healthy forests, 4–5, 23, 155, 167, 183, 205, 234–35
  Indigenous culture and, 5, 183
  logging of, for profit, 166
  mycorrhizal fungal networks and, 23, 36, 119, 163–64, 227
  natural regeneration around, 99
  need to change forest industry to conserve, 7–8
  photosynthesis and, 23
  seasonal events and, 234–35
  seed tree method and, 33, 36
  wildfires and, 11, 76–77
Mount McQuarrie, 215
Murphy-Steed, Ari, *62*, 147, 152
Murray Creek, 45
murrelet
  ancient, 110, *111*
  marbled, 171–74, 177
mushroom, 3, 91, 165
  honey, 21
  true hedgehog, 205
mycelium, 77–78
mycoheterotrophic plants, 209

mycorrhizal fungal networks
   Amanda's studies on kinship and, 29–30, 98–99
   arbuscular, and neighboring plants, 104–5
   critiques of Simard's research on, 198–99, 208–11
   ectomycorrhizal and, 50
   ectomycorrhizal vs. arbuscular mycorrhizal networks and, 101, 164, 222
   Eva's studies on cedar, yew, and maple and, 54, 104–5, 117, 159, 167
   logging and, 117–18, 126, 143, 176–77
   morels and, 78
   mother, or elder, trees as hubs for seedlings and, 36, 119, 163–64, 227
   salmon and, 91–92
   seed trees and, 150, 155
   soil food web and, 5–6, 58, 152
   symbiotic relationships with tree roots and, 4, 23
   tropical rainforests and, 221–22

NADPH, 58
Nak'adli Whut'en territory, 60
Nakusp, British Columbia, 73, 160, 198
   wildfires of 1973 and, 69–70
Narrows MTP site, 59
National Collegiate Athletic Association, 29
National Forest Inventory, 62, *140*
National Geographic, 213
native plants, 93, 166
natural selection, 32–33, 55
Nawalakw language and culture camp, 184–85, *187*
Nelson, British Columbia, 13, *16*, 32, 48, 59, 71, 74, 124, *136*, 155, 158, 194–95, 207, 213
   wildfires of 2021 and, 159
Nelson, K'odi, 183–84
Nelson, Willie, 72
nematodes, 58
New Democratic Party, 236
New York Times bestseller list, 162
New York Times Magazine, 162
New Zealand, 56

niche complementarity, 164–66, 174–75, 220
niches, vacant, in plantations, 189, 205
Nicola lakes, 161
Nicola River, 43
Nimmo Bay, 89–91
Nitinake lakes, 161
nitrogen, 18, 75, 91, 92, 165, 176, 188
Nlaka'pamux Nation, 42–46, *62*, 98, 102, 207
nontimber forest products, 180
NPR, 213
nucleic acids, 75
nucleotides, 58
nuthatch, red-breasted, 13
Nuu-chah-nulth Nation, 83, 173

oak, Portuguese, 68
oceans, 18, 20
Okanagan Lake, 125
Okanagan Landing, British Columbia, 72
Older Sister (Haida Mother Tree), *111*
old growth forest, 10, *16*, 23, 103, 108–10, 114, 119–20, 122, *129*, 135, 139–43, *140*, 148, 159–61, *162*, 163, 168, 170–76, 181, 183, 188–91, *190*, 206, 208, 211, 222, 233–35
Old Massett, British Columbia, 109–11, 120
Olympic Peninsula, 170
Onward Ranch, 26
orcas, 15
orchid, 163, 220
   fairy slipper, 5, 149, 169
Oregon State University, 125
organo-mineral compounds, 137, 164
Orrego, Gabriel, 164
otter, sea, 204
ovarian cancer, 102
overfishing, *111*
overstory, 207
   Amazon rainforest and, 220
   carbon and, 23
   cedar saplings and, 117
   ectomycorrhizal networks and, 101
   inland rainforest and, 100
   MTP sites and, 8, 33, 127–28, 145–49, 151–52, 154–55

mycorrizhal network and, 117
Pacific yew and, 103–5
understory and, 23, 117–18, 167
owl, 37
burrowing, 67
northern saw-whet, 111
western screech-owl, 174
oxygen, 18, 20, 67, 218

Pachamama Alliance, 219, 223
Pacheedaht Nation, 169, 171–72, *178, 179,* 180, 237
Pacific Rim Forest, 7, 11, 237
paclitaxel, 102–5
paintbrush, Indian, 16, 163
Pan Am Games, 31
paper industry, 68
Paris, Treaty of (1763), 22
Paris Agreements, 232
partial retention logging, 133–34, 148
Patagonia, 82, 237
pearly everlasting, 214
Penobscot Nation, 236
penstemon, 215
permafrost, 229
Perry, David, 97, 106
Peru, 218
pesticides, 93
Peter Hope Lake, *62*
pharmaceutical industry, 102–3
Phillips Arm watershed, 141–42, 176, 188, *192*
Phoenix, AZ, 79
phosphorus, 75, 91, 164, 188
phosphorylation, 58
photosynthate, 50, 100–1, 207
photosynthesis, 17, 18, 232
  ATP and NADPH and, 58
  carbon, water, oxygen and, 20
  chlorophyll and, 18
  kapok and, 223, 226
  morels and Douglas firs and, 78
  Mother Trees and, 24
  overstory vs. understory and, 23
  transpiration and, 77–79
photosynthetic spectrum, 163, 218
photosystem I, 58
photosystem II, 58

Pickles, Brian, 8, 55, 158
pine, 60, 101, 117–18
  little prince's, 174
  lodgepole, 45, 54, 56, 65, 67–68, 75, 99–100, 115, 151, 214
  ponderosa, 54, 57, 65, 67, 78, 151, 208
  radiata, 228
  western white, 162
place, importance of, 154–56, 168
plantain, rattlesnake, 50
plant roots, 58
plum, Indian, 88, 187
Pojar, Jim, 142
Port Alberni residential school, 183
Port Renfrew, 170
Portugal, 68–69, 210
predators, 182
Prince George, British Columbia, 99, 194, 201
proforestation, 208
proteins, 58, 75
Purcell Mountain Range, 59, 151
pussytoes, rosy, 152

Quatse Lake, *140*
Quebec, 205
Quito, Peru, 228
Quw'utsun Nation, 208

radiation treatments, 102
Rainbow Eyes (forest defender), 171, 177–78
rainfall, 9, 55, 77. *See also* floods
Rainforest Flying Squad, 171, 180
Ramaro (Achuar guide), 220, 223–27, *225*
raspberry, red, 149, 154
rattlesnake root, western, 174
rattlesnakes, 59
raven, common, 86, 90
Raven (Haida trickster), 229
Rayonier Canada, 108
reciprocity, 7, 9, 62, 80, 81, 103, 105, 167–68, 300
recreation, 180
reductionism, 210
regeneration, natural, 159
  MTP sites and, 147, 154

regenerative forestry, 9, 139, 141, *189*, 235
Reid, Bill, *Jade Canoe*, 229
remote sensing, 191
residential school system, 22
respiration, 20, 58
riceroot, chocolate-petaled, 87, 207
rice terraces, 88
Richardson, Miles, 109
Ridge Camp, 170–71, 173–74
Ringman, Steve, 174–75
riparian communities, 91, 205–6
Ripple Creek, 197
Rising Tide land agreement, 236
rivers and streams, 122–23, *123*
Roach, Winnifred Jean Mather, 8–9, 11, 13–14, *14*, 17, 20–26, 32–33, 42, 43, 46–48, 51–58, 61–65, 67–69, 71, 75, 78, 83, 88, 106–8, 111, 115, 119, 121, 125–27, 130–32, 134, 139, 144–47, *145*, 149–54, 198, 213, 228, 233, 237
Robinson, Alyssa, 137
Rock, Curtis, *62*
Rocky Mountains, 151
Rocky Mountain Trench, 59, 152
root gardens, 87–88, 89, 204, 235
root pathogens, 21, 165
rose, 151
    baldhip, 20, 27
Rosemont Elementary School, 74
Royal Canadian Mounted Police (RCMP), 173–74, 178–80, *179*
Russian steppe, 11
Ryan, Teresa Sm'hayetsk, 8–9, 44–47, 79–89, *81*, 92–93, 103, 107, 111, 113–15, 119–22, 155, 183–84, 186, 206, 209, 230, 232, 237

Sachs, Don, 73–74
Sachs, Hannah, 10–11, 14–15, 17, *19*, 20–21, 24–26, *62*, 72, 78, 96, 144, 150, 194, 207, 219
Sachs, Nava, 10–11, 14–15, 17, *19*, 20–21, 24–26, 53, 72–73, 75–76, 96, 144, 146, 149, 169–72, 177, *178*, 207, 213–16, 228
sagebrush, big, 67
Saint Laurence (river boat), 199

Sa ł a (Kwakwaka'wakw mourning ceremony), *187*
salal, 90, 141, 207
salamander, northwestern, 16
Salish peoples, 42
Salish Sea, 15
salmon, 79, 198
    bears, trees, and, 89–92, 175
    carbon and, 91–92, 139–40
    cedar and, 92, 175
    Haida Gwaii and, 108, *111*, *123*, 180, 206
    Indigenous people and, 8, 81–87, *82*, *87*, 108–9, 113–14, 182, 204–6, 235
    logging and, *87*, 107–8, 110, *111*, 122–23, *123*, 177, 235
    yews, 100
salmonberry, 88, 90, 141, 187, 189, 204–5
salmon species
    coho, 180
    kokanee, 198
salsify, yellow, 152
salvage logging, 38, 44, 65, 70
sandpiper, spotted, 86
San Juan Valley, 170
Sante Fe Indian Market, 79
saprotrophic fungi, 22, 78, 143, 222
sapsucker, red-breasted, 206
sapwells, 16
sarsaparilla, wild, 187
Saskatoon berry (serviceberry, juneberry), 20, 187
satellite imagery, 191
sawdust, *121*
Schacter, Teah, *140*, 146
science, 138–39
    ancient wisdom and, 105–6
    anthropomorphizing and, 209
    hard data and, 207
    limitations of, 138, 193
    reductionism and, 210–11
Scotland, 56
sea lion, 89
Seattle Times, 173
secondary forest, 208, 234
second-growth forests, 110, 140, 142, 188–91, *190*, 206, 208, 234
Secwépemc territory, 64
sedges, 113
seed banks, 206

seedlings
　burned soil and, 75
　clearcut and, 99, 151–52, 117–18, *118*, 154
　deer and, 115
　MTP sites and, 8, 54–56, 147, 152, 154
　migrant, 54–55
　mycorhizzal fungal networks and, 208–9
　rotting logs and, 5–6
　soil nutrients, and photosynthesis and respiration and, 58
　warm summers and, 115
seed tree treatments, *34*, *36*, *51*, *53*, 147–52
Sekani territory, *19*, *145*
selective harvesting, 133–34
Selkirk Mountains, 59, 123–24, 151, 213
Septuplets experiments, 100–6, 116, 157–59, 167–68
Shakespeare, William, 101
shrew, common, 16
Shuswap River, 86, *87*, 199
Silver Spray Cabin, 213–15
silverweed, yellow-flowered, 87, 207
Simard, Ellen June Gardner "Mum" "Granny Junebug," 10–11, 15–16, *16*, 21, 52–53, 71–75, 79, 112, 159–62, *162*, 193, *197*, 197–201, 216–17
　Alzheimer's and, 71–75, 95–98
　death of, 211–15, 218–19, 221, 229
　MAID and, 95–97, 159–60, 162, 197–98, 199, 201–3, 207–8
Simard, Grampa Henry, 103, 133, 199–200, 220
Simard, Granny Martha, 103, 199
Simard, Kelly, 199
　death of, 26–27, 195
Simard, Peter Ernest Charles "Dad," 102, 199–200
Simard, Robyn, 74–75, 79, 96, 112–13, 160, 197–200, 201, 203, 211, 212
Simard family, 86, *88*
Simard Mountain, 133
Simpcw Nation, 207
Sinixt Nation, *17*, *49*, *129*, *136*, *162*
Sitkum Creek fire of 2015, 73
skidders, 41, 126–27, 135, *136*, 141, 233
skunk cabbage, 79

slash pile, 49, *49*, 118–19, *118*, *121*, 132, 134, 136, 143, 233
smallpox, 204
　epidemic of 1862, 120
　outbreak of 2880s, 187
snowdrop, 73
snowfall, 9, 65–66
snowmelt, 70, 124
Snyder, Eva, 31–32, 54, 61–63, *62*, 97, 100–6, 116–17, 127, 149, 152–53, 157–59, 167–68, 172, 186–87, 196, *196*, 202, 219
soil, 8, 55
　carbon in, 6, 20, 22, 115, 176, 188
　carbon in, and logging, 126–43, 188, 232–34
　cedar, yew, and huckleberries and, 117
　deciduous trees and, 69–70
　food web and, 5, 57, 58, 77, 152, 177, 206
　grasses and, 149
　Indigenous stewardship and, 188
　organo-mineral compounds in, 6, 176
　plantations and, 189, 206
　rainfall and, 77
　salmon and, 89–91, *111*
　sampling, *140*, 141–43
　tree species access to nutrients in, 6, 164–65, 189
　water and oxygen in, 20
　whole tree logging and, 135–36
　wildfires and, 65, 75–77
soil horizons, 132, 137, 164, 176
solar eclipse, 234
soopolallie, 20–21, 151
Southern Interior Silviculture Committee (SISCO), 125, 131–39
South Moresby Wilderness Area, 108
Species at Risk Act, 174–75
Speck, Irvin, 89–91
Spences Bridge, 44–45, 155
spiders, 221
Splatsin Natin, 85–86, *87*, 200
springtails, 6, 58
spruce, 45, 60, 112
　Cariboo MTP site and, 99
　monoculture plantations, 142–43
　northern Italy and, 234
　replanting cedar clearcuts and, 117
　roots, weaving baskets and hats, 81, 88

spruce species
   interior, 22
   Sitka, 85, 88, 107, *111*, 115, 142, 185, 188–89
squirrels, 6, 67, 125, 208
   Columbian ground, 67
   red, 76
starlings, 229
St. Joseph's Mission residential school, 26–27
stomata, 17, 76–77
stonetrap fishing, *82*, 83–87
storms, 122–23
strawberry, 67, 205
   Pacific coast, 187
Stuwix Resources, 44, 47
St'uxwtéws te Secwépemc territory, *66*
subcanopy, 220
sugar metabolism, 58
sulfur, 75
Summers, Bree, 149–50
Syilx Nation, *16*, *49*, *129*, *136*
symbiotic relationships, 4, 78
synergy, 9, 166

Tahltan Nation, 180–81
tarantula, Ecuadorian purple, 221
taxane, 102
taxol, 103
Taxomyxes andreanae, 103
tea, woolly Labrador, 114
Teal-Jones Group, 169–70, 172, 180–81, 236
TED talks, 92, 163, 182–83, 219
terpenes, 77
T'exelc Nation, *34*, *35*
Tezzeron Lake, 153
thalli, 75
thimbleberry, 52, 88, 146, 153, 187, 189, 205
third-growth plantations, 188, 191
thistle, bull, 149, 152
Thompson, Douglas, 180
Thompson River, 43, 45, 64, 112
Timber Harvesting Land Base, 170, 181
timber industry. *See also* clearcutting; logging
   Amazon and, 223
   clearcutting and, *189*
   clearcutting and profits of, 98–99, 150
   climate crisis and, 40
   control of resources and, 85–86, *87*
   decline of forest and, 93
   Haida Gwaii and, 107–11, *110*, 115, *121*
   licenses taken from small-time loggers and, 43–44
   machinery and, 133
   MTP and, 8, 32
   plants surrounding buffer areas and, 56
   monoculture conifers and, 68–69
   rush to log old trees, 150
   terminology in, 22
   uplift of 1990s and profits of, 39
tissue construction, 58
Titania (mother tree), 171–72, 177, *178*, 228
Tk'emlúps te Secwépemc Nation, 31
Tl'axt'en Nation, 60
Tlielang totem pole, 120
Tlingit Nation, 83
Tod Mountain ski runs, 161
toucans, 220
tourism, 123, *123*, 180
Trans Mountain pipleline, 207
transpiration, 17, 76–78
Tree Farm License, 170, 180
tree sits, 171, 174
tropical rainforests, 222
troposphere, 218
Trust for Public Land, 236
Ts'msyen (Tsimishian) Nation, 8, 79, 81–83, *81*, 113–14
Turtle Island, 22
turtles, 220
Twaal Creek MTP site, 47, 125
twinflower, 152
Twist, Lynn, *225*
Two-Bit MTP site, 59

Umbas, Tsastilqualus Ambers "Tsas," 204–5, 207, 212
understory, 148
   carbon and, 23
   deer and, 115
   fire and, 23, 67–68, 75–76, 159
   mother trees and, 159, 172

MTP and, 8, 20
overstory and, 23, 117–18, 167
plantations and, 142, 189, 206–7
UN Intergovernmetal Panel on Climate Change, 138–39
University of British Columbia (UBC), *14*, 42, 91, 135
   Faculty of Forestry, 72
   greenhouse, 28, 157
   Indigenous Ecology Lab, 207
   Thunderbirds women's ice hockey team, 29
University of Northern British Columbia, *19*, 60
University of Victoria, 74, 204
urchins, 89

Vancouver, 8, 157, 159, 228
Vancouver International Airport, 229
Vancouver Island, 53, *140*, 141, 157, 160, 170, 183, *187*, 203, 207–8
vanilla, wild, 223
volatile organic compounds, 77
vole, red-backed, 21–22

Walkem, Dave, 42–47, 98, 133–34, 155
War in the Woods, 176
Washington State, 170
water, 16–17, 20, 67
   canopy and, 77
   cycling through forest, 76–79, 93, 229
   kapok and, 226
   old Douglas fir and saplings and, 16
   photosynthesis and, 20
   plantations and, 189
   tree diversity and, 165
Waterfall Camp, 173
water table, 154
West Kootenays, 59
Westsyde High School, 112
whisky jack, 49–50
White, Kihlguulans Kihl 'Yahda Christian, 120

White-Kuuyang, Lisa, 109–11, *111*, 114–15, 119–23, *121*, 130, 140, 156
White Queen peak, 194
Whitewater, British Columbia, *16*
Whitewater Ski Resort, 194, 203
whole tree logging, 135–37, 233
wildfire escape kits, 73
wildfires, 6, 210, 231
   deciduous trees and, 69–70
   determining risk of, *19*
   floods and, 65–66
   forest diversity and, 165
   forest floor and, 70–71
   greenhouse gases and, 231–32
   harvesting pace and, 38
   herbicides and, 205
   increase in, 27, 63–73
   Indigenous stewardship and, 7, 188
   MTP site and, 8, 10–11, 24–26, 75–76, 153
   Mum and, 15
   of 1973, 69
   of 2003, 72
   of 2017, 39, 67–73
   of 2018, 70, 73
   of 2021, 70, 159
   of 2023, 70
   old growth and, 159–61
   Portugal and, 68–69
   regeneration and, 9
   suppression efforts and, 27, 38, 68, 70, 85
   tree diversity and, 166
   whole tree logging and, 136
   zombie, 71
Williams Lake, *197*
Willie, Mike, 184, 186
willows, 16, 75, 164, 199, 216
Windy Bay watershed, 119
wintergreen, 142, 163
   pink, 50, 149
wolf, 20, 85, 153, 182
   gray, 25
wolverine, 215
Woodbury Creek, 213
woodpecker, 6, 59, 152, 165
   pileated, 16
wood ticks, 59

woollyhead, meadow, 149
Word for World Is Forest, The (Le Guin), 189
Wright, Joshua, 170

Xatśūll elder, 20–21

Yahgu'laanaas clan, 109
Yakoun River, *111*, 122
yarrow, 153, 207
Yellow Lake, 26

yew, Pacific, 54, 100–5, 116, 163, 189
   arbuscular mycorrhizal fungi and, 101
   Eva's Septuplet experiment on Douglas fir, maple, cedar, and, 103–4, 117, 157, 159, 167, 172
   MTP sites and, 149
   overstory and, 117
   paclitaxel and, 102–3
Yurok Tribe, 236

Zickmantel, Aidan, 145–46, 153

ILLUSTRATION CREDITS

Interior

Page xi: Map designed by Anneke Rosch; page 14: Blondie; page 16: Bill Heath; page 19 (*top*) Bill Heath; page 19 (*bottom*) Bill Heath; page 31: Brian Pickles; page 34 (*top*): Bill Heath; page 34 (*bottom*): Bill Heath; page 35 (*top*): Bill Heath; page 35 (*bottom*): Bill Heath; page 49: Suzanne Simard; page 62: Jean Roach; page 66: Bill Heath; page 81: Bill Heath; page 82: Bill Heath; page 87: Enderby & District Museum & Archives, EMDS 1922; page 110: Bill Heath; page 111: Bill Heath; page 118: Bill Heath; page 121: Bill Heath; page 123: Bill Heath; page 129: Deb MacKillop; page 131: Bill Heath; page 136: Bill Heath; page 140: Suzanne Simard; page 145: Bill Heath; page 162: Bill Heath; page 178: James; page 179 (*top*): Mike Graeme; page 179 (*middle*): Mike Graeme; page 179 (*bottom*): Mike Graeme; page 184: Momme Halbe; page 187: Odette Auger; page 189: Laird Miller; page 190: Danica Law; page 192: Laird Miller; page 196: Eva Snyder; page 197: Bill Heath; page 215: Teresa (*Sm'hayetsk*) Ryan; page 225: Peggy Callaghan

Insert

Page 1: Enderby & District Museum & Archives; page 2: Jean Roach; page 3: Bill Heath; page 4: Bill Heath; page 5 (*top*): Bill Heath; page 5 (*bottom*): Bill Heath; page 6: Suzanne Simard; page 7: Bill Heath; page 8: Bill Heath; page 9: Eva Synder; page 10 (*top*): Bill Heath; page 10 (*bottom*): Bill Heath; page 11: Bill Heath; page 12 (*top*): Teah Schacter; page 12 (*bottom*): Mother Tree Project; page 13 (*top*): Bill Heath; page 13 (*bottom*): Mother Tree Project; page 14 (*top*): Mike Graeme; page 14 (*bottom*): Mike Graeme; page 15: Mark Worthing; page 16: Stephanie Troughton